PROVING IN THE ELEMENTARY MATHEMATICS CLASSROOM

Proving in the Elementary Mathematics Classroom

ANDREAS J. STYLIANIDES

University of Cambridge, UK

OXFORD
UNIVERSITY PRESS

Great Clarendon Street, Oxford, OX2 6DP,
United Kingdom

Oxford University Press is a department of the University of Oxford.
It furthers the University's objective of excellence in research, scholarship,
and education by publishing worldwide. Oxford is a registered trade mark of
Oxford University Press in the UK and in certain other countries

First Edition published in 2016

Impression: 1

Published in the United States of America by Oxford University Press
198 Madison Avenue, New York, NY 10016, United States of America

British Library Cataloguing in Publication Data
Data available

Library of Congress Control Number: 2016932191

ISBN 978–0–19–872306–6

Printed and bound by
CPI Group (UK) Ltd, Croydon, CR0 4YY

To Lesia and Yannis

FOREWORD

Toward the end of the twentieth century, as numerous studies revealed the difficulties for students in moving from arithmetic to algebra (cf. Kieran, 1992; Kieran, Pang, Schifter, & Ng, 2016; Wagner & Kieran, 1989), the question arose, What can be done in the elementary grades to better prepare students for the transition? During the subsequent decades, a number of research teams pursued this question, particularly investigating students' engagement with activities involving functions (Blanton, 2008; Carraher, Schliemann, Brizuela, & Earnest, 2006; Malara & Navarra, 2002; Moss & London McNab, 2011; Radford, 2014) and generalized arithmetic (Britt & Irwin, 2011; Carpenter, Franke, & Levi, 2003; Russell, Schifter, & Bastable, 2011a; Schifter, Monk, Russell, & Bastable, 2008).

The goal of much of this early algebra work has been to promote a way of thinking—the habit of looking for regularity, and articulating, testing and proving rules or conjectures (Kieran et al., 2016). These studies all shared an emphasis on students' reasoning rather than a more limited focus on fluent use of calculation procedures. Through classroom interaction in which students elaborate their own thinking and engage with their classmates' ideas, they consider, evaluate, challenge, and justify hypotheses thus participating in proving activity.

But what does it mean for elementary-aged students, who do not have access to formal mathematical tools for proof, to engage in proving mathematical claims? Do young students' proving activities constitute proofs? In what ways? As these young students engage in mathematical reasoning, how does that activity connect to and prepare them for understanding proof in more advanced mathematics, a challenging topic even for older students and adults (Harel & Sowder, 1998; Knuth, 2002a)? While there has been a strong push in various policy documents for the inclusion of mathematical argument throughout students' schooling (cf. National Governors Association for Best Practices & Council of Chief State School Officers [NGA & CCSSO], 2010), these documents are generally thin in providing the characteristics of proof, types of proving activities, and examples of what this work might look like and require of teachers in the elementary grades.

In his book, *Proving in the Elementary Mathematics Classroom,* Andreas Stylianides has contributed a substantial resource to this discussion. Rich with images of 8–9-year-olds working together on mathematics problems, the book analyzes students' words and actions in terms of what one should look for and expect of young children engaged in proving activities. Examples illustrate how teachers set up proving tasks and interact with their students to challenge their thinking and move them toward proof. Furthermore, the book demonstrates how proof and proving can be integrated into the study of numerical calculation, the heart of mathematics content in the elementary grades: How many two addend expressions can be made with the sum of 10? What possible numbers can be made with the digits 1, 7, and 9? When is a product greater than its factors? Questions such as these provide fertile ground for proving.

Stylianides offers a definition of proof in the context of a classroom community that includes three criteria:

> *Proof* is a *mathematical argument*, a connected sequence of assertions for or against a mathematical claim, with the following characteristics:
>
> 1. It uses statements accepted by the classroom community (*set of accepted statements*) that are true and available without further justification;
> 2. It employs forms of reasoning (*modes of argumentation*) that are valid and known to, or within the conceptual reach of, the classroom community; and
> 3. It is communicated with forms of expression (*modes of argument representation*) that are appropriate and known to, or within the conceptual reach of, the classroom community. (Stylianides, 2007b, p. 291) (emphasis in original)

One of the key contributions of this book is the analysis of classroom examples with respect to these three aspects of proof. Modes of argumentation, Stylianides makes clear, are not only within the conceptual reach of the classroom community, but are also consistent with those that are accepted by mathematicians as proof. For example, providing numerous examples in support of a conjecture that covers infinitely many cases does not constitute proof, while refutation of a conjecture is established with one counterexample.

A second important contribution is Stylianides' taxonomy of proving tasks based on the cardinality of the question—proof involving a single case, multiple but finitely many cases, or infinitely many cases—and whether a conjecture is being proved true or false. Each category lends itself to a different form of argument. This taxonomy thus has the potential to inform researchers, teachers, teacher educators, and curriculum developers about the range of proving activities to be considered.

In this book, each type of task is illustrated through classroom examples, providing a structure for understanding the territory of proving at these grade levels. Proving tasks take students deeper into the underlying mathematics of the content under study, even when the task involves only a single case. For example, in Chapter 6, the teacher posed a combinatorics problem—how many outfits can be made with three dresses and two hats?—and then clarified that she wanted students to say "something interesting, not the actual answer." Moving beyond a single-minded focus on "the actual answer," students come to recognize that looking at mathematical patterns and regularities often leads to "something interesting," something about mathematical structure and relationships.

We, the authors of this Foreword, have found in our work that students who are given regular opportunities to notice patterns across related problems, are encouraged to articulate what those patterns are, and are asked to develop arguments about why they occur, become attuned to looking for regularities in mathematics (cf. Russell et al., 2011a). Without prompting from the teacher, they come up with ideas about what might be true and offer their own conjectures. That is, they become curious about how mathematics works and develop tools they need to test and prove their ideas. As Stylianides points out in the concluding chapter, "once classroom norms that support argumentation and proof have been established, students themselves can also raise the issue of proof and can engage in proving activity independently of the teacher's presence" (p. 159)

This book demonstrates that work on proof and proving can engage the whole range of students in significant mathematical reasoning, from those who have a history of struggling with school mathematics to those who have excelled. As one of our collaborating teachers said recently,

> [When working to notice, articulate, and prove generalizations about the operations], there are so many opportunities for the struggling students to continue to work on their ideas, and at the same time the more advanced students can continue their work, pushing themselves to think further about a particular concept, representation, conjecture, etc. (Russell, Schifter, Bastable, Higgins, & Kasman, in press)

Stylianides points out that recent research indicates that teachers can learn content and teaching practices that support mathematical argument and proof in elementary classrooms. In our work with a range of classroom teachers over the past decade, we, too, have found that teachers can learn the relevant mathematics content, learn how young students engage in proving, and learn teaching practices that support students in this realm (Russell, Schifter, Bastable, & Franke, submitted). However, most teachers receive their teaching certification ill prepared to include proof and proving activities in their instruction. Many teachers at the beginning of their work with us report that they do not have experience supporting mathematical argument in their classroom and, in fact, are not sure what mathematical proof is, or what it can be for young students. This is an area teacher education programs have yet to take on. Thus, *Proving in the Elementary Mathematics Classroom* provides an important resource for researchers to continue investigations into proof and proving in the elementary grades, for teachers to develop images of students engaged in proving activities, and for teacher educators to help practicing and prospective teachers bring proving into their classrooms.

Deborah Schifter, Principal Research Scientist
Education Development Center, Inc., Waltham, MA, USA

Susan Jo Russell, Principal Research Scientist
Education Research Collaborative at TERC, Cambridge, MA, USA

PREFACE AND ACKNOWLEDGMENTS

In *Proving in the Elementary Mathematics Classroom* I address a fundamental problem in children's learning that has received relatively little research attention thus far: Although proving is at the core of mathematics as a sense-making activity, it currently has a marginal place in elementary mathematics classrooms internationally. My broad aim in this book is to offer insights into how the place of proving in elementary students' mathematical work can be elevated. In pursuing this aim, I focus on mathematics tasks, which have a major impact on the work that takes place in mathematics classrooms at the elementary school and beyond.

Specifically, I examine different kinds of proving tasks and the proving activity that each of them can help generate during its implementation in the elementary classroom. I examine further the role of elementary teachers in mediating the relationship between proving tasks and proving activity, including major mathematical and pedagogical issues that can arise for teachers as they implement each kind of proving task. My examination is situated in the context of classroom episodes that involved mathematical work related to proving in two different elementary classes: a Year 4 class in England (8–9-year-olds) taught by a teacher whom I call Mrs. Howard (pseudonym[1]), and a third-grade class in the United States (again 8–9-year-olds) taught by Deborah Ball. I also studied Ball's teaching practice in my prior research on proving in elementary school mathematics (Stylianides, 2005, 2007a–c; Stylianides & Ball, 2008); this book is partly a synthesis of that work, through the particular lens of proving tasks and proving activity as mediated by the teacher. Most importantly, though, this book is an extension and further development of that work based on an expanded corpus of classroom data. This includes the data from Mrs. Howard's class, which I collected specifically for the book and the analysis of which has offered fresh (mostly complementary) insights into the issues of interest. Of course, the expansion and further development of my earlier work in this book has also benefited from major developments in the relevant research literature.

This book makes a contribution to research knowledge in the intersection of two important and related areas of scholarly work, with a focus on the elementary school level: the teaching and learning of proving, and task design and implementation. In addition to this contribution, the work reported in this book has important implications for teaching, curricular resources, and teacher education. For example, it informs the use of specific proving tasks in the service of specific learning goals in the elementary classroom. Also, this book identifies different kinds of proving tasks whose balanced representation in curricular resources can support a rounded set of learning experiences for elementary students related to proving. It identifies further important mathematical ideas and pedagogical practices

[1] Under the provisions of the permission I received to conduct research in Howard's school, I am obligated to refer to her using a pseudonym.

related to proving that can constitute objects of study in teacher education programs for pre-service or in-service elementary teachers.

Many people supported my efforts in the development of the work reported in this book, and I am indebted to all of them. I owe a tremendous debt of gratitude to Deborah Ball, Mrs. Howard, and another English teacher, whom I call Mrs. Lester (also a pseudonym[2]), for allowing me to study their teaching practices. Ball should be credited also as the person who instilled in me a strong research interest in elementary school mathematics and helped deepen my thinking about the topic at the early stages of my research journey. Howard and Lester allowed me access not only to their classes but also to their professional thinking, and they graciously accommodated my constraints for the school visits. My visits were inspiring, and each I time I left their schools I felt reassured that this project was worth undertaking. For reasons that I explain in Chapter 3, I made the difficult decision not to include an analysis of Lester's practice in the book. This was unfortunate as Lester's practice could be of broader interest, like that of Ball and of Howard.

Over the years, in the different places where I worked or studied, I have been fortunate to receive critical commentaries from, or engage in friendly discussions with, a number of individuals who have benefitted my thinking about ideas related to the book. This was a diverse group of researchers and practitioners and included the following: Hyman Bass, Seán Delaney, Kim Hambley, Patricio Herbst, Mark Hoover, Magdalene Lampert, Judith Large, Zsolt Lavicza, Adam Lefstein, John Mason, Raven McCrory, Karen Russell, Susan Jo Russell, Kenneth Ruthven, Geoffrey Saxe, Deborah Schifter, Alan Schoenfeld, Helen Siedel, Edward Sylver, and Hannah Waterhouse. I thank all of them. Special thanks go to Gabriel Stylianides, my close collaborator and brother, for his time, insights, and valuable help with overcoming hurdles related to conducting the research and writing the book.

I also thank my editor, Keith Mansfield, and three anonymous reviewers whose critical questions and comments led to a better book, as well as Dan Taber and other Oxford University Press staff for their helpful assistance in preparing the book for publication. I am also grateful to Deborah Schifter and Susan Jo Russell for kindly agreeing to write the Foreword.

Last, but not least, I express my gratitude to my wife, Lesia, not only for her continuous support but also for her questions and comments that challenged my thinking, and to our son, Yannis, for being a constant source of inspiration for me.

[2] I am obligated to refer to Lester using a pseudonym for the same reason as for Howard (see note 1).

CONTENTS

1 **Introduction** 1
A Classroom Episode of Elementary Students Engaging With Proving 1
Description of the Episode 2
Brief Discussion of the Episode 3
The Book's Broad Aim and Intended Audience 4
The Book's Structure 6

2 **The Importance and Meaning of Proving, and the Role of Mathematics Tasks** 7
The Importance of Proving 7
A Philosophical Argument 8
A Pedagogical Argument 9
The Meaning of Proving 10
An Overview of the Situation in the Field 11
The Definition of Proving Used in this Book 11
The Definition of Proof Used in this Book 13
The Place of Proving in Mathematics Classrooms 19
The Current Place of Proving 20
The Role of Mathematics Tasks in Elevating the Place of Proving 24

3 **The Set-up of the Investigation** 27
A Categorization of Proving Tasks 27
Characteristics of Proving Tasks That Can Influence Their Generated Proving Activity 32
Data Sources and Analytic Method 35
Data Sources 35
Analytic Method 39

4 **Proving Tasks with Ambiguous Conditions** 41
Episode A 41
Description of the Episode 41
Discussion of the Episode 52
Episode B 58
Description of the Episode 58
Discussion of the Episode 64
General Discussion 68
The Relationship between Proving Tasks and Proving Activity 68
The Role of the Teacher 70

5 Proving Tasks Involving a Single Case 73
Episode C 74
Description of the Episode 74
Discussion of the Episode 79
Episode D 81
Description of the Episode 81
Discussion of the Episode 83
General Discussion 86
The Nexus between Calculation Work and Proving 86
The Relationship between Proving Tasks and Proving Activity 87
The Role of the Teacher 88

6 Proving Tasks Involving Multiple but Finitely Many Cases 91
Episode E 91
Description of the Episode 91
Discussion of the Episode 94
Episode F 97
Description of the Episode 97
Discussion of the Episode 112
General Discussion 118
The Relationship between Proving Tasks and Proving Activity 118
The Role of the Teacher 120

7 Proving Tasks Involving Infinitely Many Cases 123
Episode G 123
Description of the Episode 123
Discussion of the Episode 132
Episode H 136
Description of the Episode 136
Discussion of the Episode 143
General Discussion 148
The Relationship between Proving Tasks and Proving Activity 148
The Role of the Teacher 150

8 Conclusion 153
The Relationship between Proving Tasks and Proving Activity 153
The Role of the Teacher 157
Selecting or Designing Proving Tasks 158
Implementing Proving Tasks 158
The Place of Proving in Elementary Students' Mathematical Work 162
Teacher Education 163
Curricular Resources 164
Epilogue 166

References 167
Author Index 181
Subject Index 185

1

Introduction

When people think about the notion of "proof" in mathematics, they usually have in mind advanced mathematics courses at secondary school or university level. This is not surprising. In many countries around the world the notion of proof and the corresponding activity of proving have traditionally been part of the mathematical experiences of only advanced secondary students or university students majoring in mathematics, with elementary students being offered limited (if any) opportunities to engage with proving. However, and as I shall explain in more detail in Chapter 2, the marginal place of proving in elementary school mathematics is seriously problematic. There are at least two reasons for that. First, elementary students are deprived of opportunities to learn deeply in mathematics. Second, when students ultimately encounter proving at secondary school or university it feels alien to them rather than a natural extension of their earlier mathematical experiences at elementary school.

This book challenges the popular perception of proving as being an advanced mathematical topic beyond the reach of elementary students, and examines what it might take to elevate the place of proving in elementary students' mathematical work. But before I say more about what the book is about, whom it can be for, and how it is structured, I consider briefly an episode from a fifth-grade class to exemplify what it looks like when elementary students engage with proving.

A Classroom Episode of Elementary Students Engaging With Proving

This episode took place in a Canadian fifth-grade class (10–11-year-olds) taught by Vicki Zack, a teacher–researcher. The description of the episode derives from one of Zack's publications (Zack, 1997; see also Zack, 1999) and is limited to a summary of few classroom events that occurred over different lessons. While the episode is not illustrative of what typically happens in elementary classrooms nowadays, it nevertheless illustrates what might be possible in such classrooms.

Description of the Episode

The students in the class were working on a mathematics task, which I will call the Squares Problem. The problem engaged students in finding (an expression for) the total number of different squares in tiled squares of given sizes, and in proving that indeed these are all the different squares there are. Consider, for example, the 4-by-4 square in Figure 1.1. This 4-by-4 square has 16 "little squares" (i.e., 1-by-1 squares); a smaller number of 2-by-2 squares (how many?); an even smaller number of 3-by-3 squares (how many?); and finally one "big square" (i.e., the 4-by-4 square itself). How many squares are there altogether?

The problem came in four parts that students examined in order. Part 1 asked students to find all the different squares in a 4-by-4 square like the one in Figure 1.1, and to prove that they have found them all. Parts 2–4 asked students to find and prove how many squares there would be if, instead of a 4-by-4 square, they were given, respectively, a 5-by-5 square, a 10-by-10 square, or a 60-by-60 square.

The students explored the Squares Problem first in small groups and then in slightly larger groups. A group of three, with Gord, Lew, and Will, joined a group of two, with Ross and Ted, to form a group of five. The students followed a variety of approaches to parts 1–3 of the problem, but all of them came up with the same answers: 30 for part 1; 55 for part 2; and 385 for part 3.

For example, for part 1 Lew and Ross used variants of a "trace and count" approach, with Lew in particular sliding each size square across the master grid and counting the number of squares of that size. Will had an intuition from the beginning that there was a pattern to be discovered and, indeed, after exploring part 1 he came up with the pattern presented in Figure 1.2, which Zack called the "criss-cross pattern." This visual method of criss-crossing gave Will an easy way to find the number of different squares of each size in a 4-by-4 square (part 1). Will tested the pattern on the 5-by-5 square (part 2) and found out that it worked again. He then assumed that the pattern would continue to work in the same way for tiled squares of different sizes. According to Zack, Will was "subsequently seen to act in

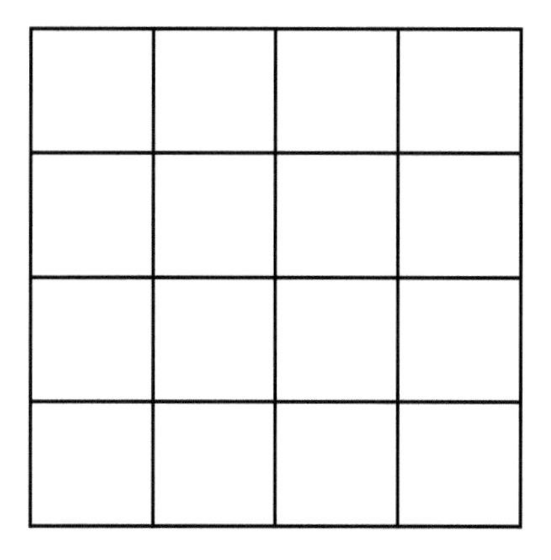

Figure 1.1 A 4-by-4 square.

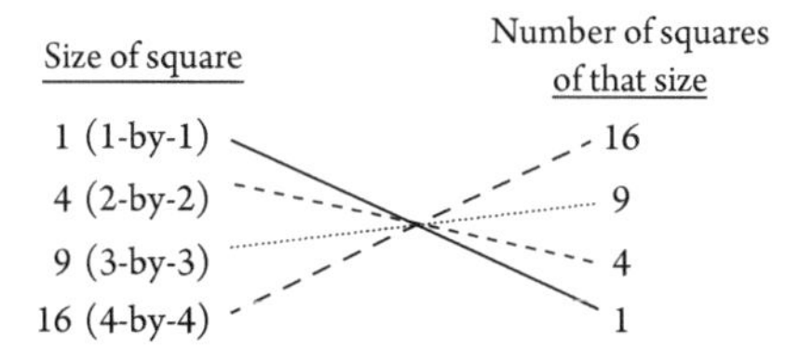

Figure 1.2 Will's criss-cross pattern in the case of a 4-by-4 square (adapted from Zack, 1997, p. 294).

accordance with this assumption" (Zack, 1997, p. 294). Will's criss-cross pattern made an impression on Gord and Lew and influenced their work. For example, it was Gord, not Will, who saw that the numbers in the second column of Figure 1.2 represented square numbers. Gord and Lew ultimately identified a pattern whereby the number of different squares in a tiled square of a given side length, say of size 5, was given by the sum of all the square numbers up to that size; in the particular case of size 5, the answer would be given by the expression $1^2 + 2^2 + 3^2 + 4^2 + 5^2$. Gord and Lew were thrilled with the discovery of this pattern and the outcomes of their collaborative work, and they conjectured that the answer for part 4 would be given by the expression $1^2 + 2^2 + 3^2 + \ldots + 59^2 + 60^2$.

When the group of five started discussing part 4, Ross and Ted claimed that the total number of squares in a 60-by-60 square would be obtained by taking the total number of squares in a 10-by-10 square (i.e., 385) and multiplying that number by 6. The answer to part 4, therefore, would be 2310. While 10 out of a total of 26 students in the class used the same method in part 4 as Ross and Ted did, the method is actually invalid and the corresponding answer is false. Gord, Lew, and Will were very certain about that, and explicitly questioned Ross and Ted's claim. This is illustrated in the following dialogue between Lew (L), Will (W), and Ross (R) (Zack, 1997, p. 295; emphasis added):

L: I'll make you a bet [that you are wrong].
W: I'll make you a bet.
L: I'll bet you anything in the world.
R: I'm not betting. *You have to prove us wrong.*

Gord, Lew, and Will took up the challenge set before them by Ross ("You have to prove us wrong"), and they offered three arguments aiming to refute Ross and Ted's claim. One of these arguments, by Lew and Will, went as follows: Even if one considered only the "little squares" (i.e., the 1-by-1 squares) in a 60-by-60 square, their number would be $60 \times 60 = 3600$, which is already bigger than 2310. In the end, Ross and Ted were persuaded that it was impossible for the answer to part 4 to be 2310. Ted said: "Yeah, they're-, you guys are right, I go along with you guys" (Zack, 1997, p. 295).

Brief Discussion of the Episode

This episode showed elementary students debating about the truth of a mathematical claim and resolving their disagreement not by means of the authority of the teacher but rather based on *proof*, using reason and mathematical argument. In doing so, the students engaged in mathematics as a sense-making activity: instead of memorizing given facts or procedures, they explored patterns and sought to understand the mathematical grounds on which claims could be accepted or rejected.

At the center of students' mathematical work was the Squares Problem, a rich mathematics task that helped create a powerful learning environment for students, including opportunities for them to make different claims and generate arguments to justify or refute those claims. Take, for example, Ross and Ted's claim that the total number of different squares in a 60-by-60 square could be obtained by taking the total number of different squares in a 10-by-10 square (i.e., 385) and multiplying that number by 6. This claim was a *false* statement about the specific case of a 60-by-60 square, and attracted three arguments, all of which aimed to refute it. Other claims made or alluded to by students in the episode included *true* statements such as the following: there are 30 different squares in the specific case of a 4-by-4 square, and the total number of different squares in the general case of an n-by-n square are given by the expression $1^2 + 2^2 + 3^2 + \ldots + n^2$.

The episode raises a number of questions: Was Lew and Will's argument, which aimed to refute Ross and Ted's false claim, a proof? What could be possible proofs that would aim to justify the true claims mentioned earlier? What mathematical resources would be required of students for the development of each of those proofs, and how might those resources or proofs vary depending on the particular kind of claim they were associated with? Would those resources be within students' reach? What was involved for Zack as she implemented in her class this rich mathematics task, and what role did she play in managing her students' proving activity? These questions illustrate more general issues that I shall address in the book and about which the field of mathematics education knows relatively little.

The Book's Broad Aim and Intended Audience

In the past few decades, proving has received considerable attention internationally. There has been an upsurge of research articles and specialized books or edited volumes on various aspects of proving (e.g., Hanna & de Villiers, 2012; Hanna, Jahnke, & Pulte, 2010; Reid & Knipping, 2010; Stylianou, Blanton, & Knuth, 2009), as well as a number of national curriculum frameworks in different countries calling for an important role for proving in the mathematical experiences of all students and from the beginning of their schooling (e.g., Department for Education, 2013; NCTM, 2000). Furthermore, major conferences in the field, such as the International Congress on Mathematical Education (ICME) and the Congress of European Research in Mathematics Education (CERME), have well-established Working Groups or Topic Study Groups that specifically aim to review, examine, and support research developments related to proving in mathematics education.

However, despite this significant interest in proving and the many researcher and curricular calls for proving to play an important role in school mathematics as early as elementary school, proving tends to have a marginal place in elementary students' mathematical work. Also, relatively little research thus far has explored possible ways to address this problematic situation. In this book, I take a step toward addressing this gap in the body of research, aiming to offer insights into what it might take to elevate the place of proving in elementary students' mathematical work, thus also enhancing the access of elementary students to mathematics as a sense-making activity.

To address this broad aim, I focus on mathematics tasks which have a major impact on the work that takes place in mathematics classrooms at the elementary school and beyond (e.g., Boston & Smith, 2009; Christiansen & Walther, 1986; Sears & Chávez, 2014). Specifically, I focus on the design and implementation of a special category of mathematically rich and cognitively demanding tasks, namely *proving tasks*, and I examine different kinds of proving tasks and the proving activity that each of them can help generate during its implementation in the elementary classroom. I examine further the role that elementary teachers play, or can play, in mediating the relationship between proving tasks and proving activity, including major mathematical and pedagogical issues that can arise for teachers as they implement each kind of proving task. My work is situated in the context of classroom episodes that involved mathematical work related to proving and offer vivid images of elementary students engaging in proving, with support from their teachers. In Chapters 1–3 I use a few episodes from published reports to illustrate key ideas, as I did with the episode from Zack's class in this chapter, but in the following chapters, which form the bulk of the book, I use episodes from two elementary classes where I conducted research.

The book is (1) a synthesis of part of my prior research on proving in elementary school mathematics (Stylianides, 2005, 2007a–c; Stylianides & Ball, 2008), through the particular lens of proving tasks and corresponding proving activity as mediated by the teacher, and (2) an extension and further development of that work. With regard to the latter, I use an expanded corpus of classroom data that includes new data that I collected specifically for the book, the analysis of which has offered fresh (mostly complementary) insights into the issues of interest. Of course, the expansion and further development of my earlier work in this book has also benefited from major developments in the relevant research literature. Some of these developments have been discussed in two recent and rather comprehensive reviews of the state of the art (Stylianides, Bieda, & Morselli, 2016a; Stylianides, Stylianides, & Weber, 2016b).

Although in writing this book I have adopted the American terminology for the level of education to which the book refers (elementary school) and for learners at that level (elementary students), I believe that the book will be of interest to a diverse audience internationally. Due to its focus on an important but under-researched topic, the book can be of interest to researchers and graduate students in mathematics education, especially those working in the areas of elementary school mathematics, proof and proving, task design and implementation, or classroom-based practices. Due to the issues it addresses pertaining to the teaching and learning of a significant mathematical activity (i.e., proving) that is currently neglected in elementary mathematics classrooms, the book can also be of interest to teacher educators, especially those who aim to prepare elementary teachers for "reform-oriented" teaching. Due to the issues it addresses pertaining to task design and implementation in the area of proving, which has traditionally received limited attention in curricular resources at the elementary school level and beyond, the book can appeal further to curriculum developers, including textbook authors. Finally, due to the way it blends a research perspective with the study of practical issues related to the teaching and learning of elementary school mathematics, the book (or individual chapters thereof) can be of interest to pre-service or in-service elementary teachers in teacher education programs, especially those who specialize in mathematics.

The Book's Structure

The book is organized into three parts. Part I comprises Chapters 1–3 and offers the background (motivation, research basis, conceptual/theoretical underpinning, etc.) for the rest of the book. In Chapter 2 I elaborate on the importance of proving for elementary students' learning of mathematics, and I clarify what I mean by "proof" and "proving" in the book. I also discuss the marginal place that proving has in typical classrooms nowadays, and I argue that proving tasks offer a promising way to elevate the place of proving in elementary students' mathematical work. In Chapter 3 I set up the investigation that I undertake in subsequent chapters: I describe a categorization of proving tasks and identify main characteristics of these categories that can influence proving activity in the classroom. Also, I describe the data sources and analytic method, and I explain my rationale for restricting my examination in the book to classroom episodes from only two out of the four classes I could have studied. These were a third-grade class in the United States (8–9-year-olds) and a Year 4 class in England (again 8–9-year-olds); the students' ages in the two classes roughly correspond to the middle age band in elementary schools internationally.

Part II is the bulk of the book and comprises Chapters 4–7. In each of these chapters I examine a different category of proving tasks and the proving activity that these tasks can help generate in the classroom, considering also the role of the teacher. In each chapter my examination is situated in two classroom episodes, one from each of the two classes just mentioned. All four chapters have a similar structure: they begin with a separate description and discussion of each episode, followed by a more general discussion. The following two issues, which reflect the specific goals of the book, are addressed in the general discussion of every chapter: (1) the relationship between the particular category of proving tasks examined in the chapter and the corresponding proving activity; and (2) the role of the teacher during the classroom implementation of proving tasks belonging to the particular category. Directions for future research are also considered in the general discussion of each chapter.

Part III comprises Chapter 8, the final one in the book. In this chapter I draw some general conclusions about the book's two core issues in light of my discussion in the previous chapters, notably the four chapters in Part II. I also identify few major directions for future research, which supplement those identified in previous chapters, and I revisit the book's broad aim to discuss implications of the work reported herein for what it might take to elevate the place of proving in elementary students' mathematical work.

I hope that the book will be read in its entirety, but I recognize that its length may not allow that, especially if it is used as a reading in teacher education programs. In such cases I would suggest referring to some necessary background from Part I, and using that alongside any, and as many as possible, of the chapters in Part II. Indeed, although there is a rationale for the particular sequencing of the chapters in Part II, and also some cross-referencing between chapters, these chapters could nevertheless be read independently of each other and in any order.

2

The Importance and Meaning of Proving, and the Role of Mathematics Tasks

Proving currently has a marginal place in typical mathematics classrooms around the world, especially at elementary school. Yet it is recognized that proving deserves an important place in students' mathematical experiences from the start of their education. In this chapter I review relevant literature to justify the importance of proving as early as elementary school, and I clarify the meanings of "proof" and "proving" as used in the book. I also discuss factors that might have contributed to the marginal place of proving at elementary school, and I argue that mathematics tasks can play a significant role in efforts to elevate the place of proving in the mathematical work of students at elementary school and beyond.

The Importance of Proving

Proving has a pivotal role in the field of mathematics. As Schoenfeld (2009) stated vividly: "If problem solving is the 'heart of mathematics,' then proof is its soul" (p. xii). For example, from an epistemic standpoint, "a proof for mathematicians involves thinking about new situations, focusing on significant aspects, using previous knowledge to put new ideas together in new ways, consider relationships, make conjectures, formulate definitions as necessary and to build a valid argument" (Tall et al., 2012, p. 15). One cannot deny that there is some debate about the place and nature of proof within the field of mathematics (e.g., Lakatos, 1976), but the point nevertheless stands that proving is indispensable to the work of mathematicians and to their efforts to deepen mathematical understanding (e.g., Davis & Hersh, 1981; Kitcher, 1984; Polya, 1981; Schoenfeld, 2009).

Although proving has a pivotal role in the field of mathematics, its place in school mathematics was for a long time unclear and its suitability for school students often debated (see, e.g., Hanna & Jahnke, 1996; Steen, 1999; Stylianides, 2008c). The general trend was that only "advanced" secondary students were offered opportunities to engage with proving, and this engagement was typically done in the context of courses on Euclidean geometry. In the past few decades, however, many researchers in mathematics education and curriculum frameworks internationally have criticized this restricted view of proving in school mathematics and called for it to become a central part of the mathematical experiences of all students across all mathematical domains (not just geometry) and at all levels of education (e.g., Ball & Bass, 2003; Ball, Hoyles, Jahnke, & Movshovitz-Hadar, 2002; Department for Education, 2013; Hanna, 1995, 2000; Hanna & Jahnke, 1996; National Council of Teachers of Mathematics [NCTM], 2000; National Governors Association Center for Best Practices & Council of Chief State School Officers [NGA & CCSSO], 2010; Stylianou et al., 2009; Yackel & Hanna, 2003). In other words, the current research and policy discourse is not simply in favor of a significant presence of proving in school mathematics but is also asking boldly for its incorporation into the mathematical experiences of even elementary school students.

I am not going to go into details about the historical evolution of the (recommended) place of proving in school mathematics (a relevant discussion can be found in Stylianides, 2008c). Rather, I will exemplify a small part of this evolution by reference to the different status that proving had in two curriculum frameworks published by the National Council of Teachers of Mathematics (NCTM, 1989, 2000) in the United States. While the older framework (NCTM, 1989, p. 143) recommended that proving be part of the mathematical experiences of *only* those high-school students intending to go to college, the more recent one (NCTM, 2000, p. 56) recommended that *all* students of *all* ages should be given opportunities to "recognize reasoning and proof as fundamental aspects of mathematics; make and investigate mathematical conjectures; develop and evaluate mathematical arguments and proofs; select and use various types of reasoning and methods of proof." The NTCM (2000) recommendations were reaffirmed by subsequent policy documents in the United States (e.g., NGA & CCSSO, 2010). Similar recommendations were made in policy documents in other countries, including England. In particular, and contrary to its predecessor, the recently published English national curriculum for mathematics (Department for Education, 2013) has set for students of all ages a core aim specifically related to proving: "[All students should] reason mathematically by following a line of enquiry, conjecturing relationships and generalisations, and developing an argument, justification or proof using mathematical language" (Department for Education, 2013, p. 3). Interestingly, this is one out of only three core aims in mathematics set by the English national curriculum for students at all levels of education (elementary and secondary).

The recommendations of researchers and curriculum frameworks that I have just summarized offer face validity to the view that proving deserves an important place in students' mathematical experiences as early as elementary school. But there are also more substantial arguments for the validity of this view. Next I discuss two related arguments, which I present separately for the sake of clarity: one philosophical and the other pedagogical.

A Philosophical Argument

From a philosophical standpoint it can be argued that proving deserves a central place throughout the school mathematics curriculum because of its central place in the field of mathematics. There are two pillars to this argument. The first is the pivotal role of proving in the field of mathematics, as outlined earlier. The second relates to theoretical ideas expressed by educational scholars such as Bruner (1960) and Schwab (1978) that the school curriculum in any subject area should be, from the start of students' schooling, an undistorted representation of the structure of the respective discipline.

According to Bruner (1960), there should be "a continuity between what a scholar does on the forefront of his [or her] discipline and what a child does in approaching it for the first time" (pp. 27–28). Similarly, Schwab (1978) argued for a school curriculum "in which there is, from the start, a representation of the discipline" (p. 269) and in which students have progressively more intensive encounters with the inquiry and ideas of the discipline. Schwab emphasized further the serious consequences that can result from a possible misalignment between the structure of a discipline and students' experiences with the respective subject area at school:

> *How* we teach will determine *what* our students learn. If a structure of teaching and learning is alien to the structure of what we propose to teach, the outcome will inevitably be a corruption of that content. And we will know that it is.
>
> Schwab (1978, p. 242) (emphasis in original)

Thus, limited attention to proving in students' mathematical education, even at elementary school, can be a serious threat to the integrity of their opportunities to learn mathematics. The idea is not that instruction should treat students as "little mathematicians," and indeed no one denies that "[c]hildren are different than mathematicians in their experiences, immediate ambitions, cognitive processing power, representational tools, and so on" (Hiebert et al., 1996, p. 19). Rather, the idea is that, similarly to its role in the field of mathematics, proof is indispensable to engagement with "authentic mathematics" at school as well (e.g., Lampert, 1992; Stylianides, 2007a). According to Lampert (1992), "[c]lassroom discourse in 'authentic mathematics' has to bounce back and forth between being authentic (that is, meaningful and important) to the immediate participants and being authentic in its reflection of a wider mathematical culture [where proof has a pivotal role]" (p. 310).

A Pedagogical Argument

From a pedagogical standpoint it can be argued that engagement with proving is necessary for deep learning in mathematics (e.g., Ball & Bass, 2003; Hanna, 1990, 1995; Mason, 1982; Yackel & Hanna, 2003). It would appear contradictory to talk about an emphasis on understanding in school mathematics, as this is nowadays the case in curriculum frameworks internationally (e.g., Department for Education, 2013; NCTM, 2000), without also paying serious attention to the role that proving can play in students' mathematical work. In particular, and as illustrated in the episode from Zack's (1997) fifth-grade class that I described in Chapter 1, engagement with proving can allow students to participate in mathematics as a sense-making activity: explore why things "work" in mathematics rather than simply memorize given facts or procedures; reconcile their disagreements or resolve debates about the truth of mathematical assertions by means of the logical structure of the mathematical system rather than by appeal to the authority of the teacher or the textbook; and overall become more active participants in knowledge construction rather than passive recipients of pre-packaged knowledge transmitted to them by the teacher (e.g., Ball & Bass, 2000b; Carpenter, Franke, & Levi, 2003; Hanna & Jahnke, 1996; Lampert, 1990, 1992;

Lin & Tsai, 2012; Maher & Martino, 1996; Reid, 2002; Wood, 1999; Yackel & Cobb, 1996; Zack, 1997).

Currently there is limited attention to proving in students' mathematical experiences, especially at elementary school. According to a group of researchers from four different countries (the United States, England, Germany, and Israel), mathematics instruction at elementary school tends to focus "on arithmetic concepts, calculations, and algorithms, and, then, as [students] enter secondary school, [students] are suddenly required to understand and write proofs, mostly in geometry" (Ball et al., 2002, pp. 907–908). In other words, the transition of students from elementary to secondary school mathematics tends to be abrupt and often represents a "didactical break" with the introduction of a new requirement for proving (Balacheff, 1988a; Sowder & Harel, 1998). As a result, at secondary school, when many students engage with proving for the first time, it feels like an alien activity to them rather than a natural extension of their earlier mathematical experiences at elementary school. Furthermore, this abrupt transition to proving at secondary school has been discussed in the literature (e.g., Mariotti, 2000; Moore, 1994; Sowder & Harel, 1998; Stylianides, 2007c; Stylianides et al., 2016b; Usiskin, 1987) as one of the reasons for the difficulties that even advanced secondary students face with proof (e.g., Buchbinder & Zaslavsky, 2007; Chazan, 1993; Coe & Ruthven, 1994; Healy & Hoyles, 2000; Hoyles & Küchemann, 2002; Knuth, Choppin, Slaughter, & Sutherland, 2002; Küchemann & Hoyles, 2001–2003; Senk, 1989; Yu, Chin, & Lin, 2004).

To summarize, from a pedagogical standpoint a more central place of proving in the mathematical experiences of elementary students can help achieve two important goals. First, elementary students will have more opportunities to learn deeply in mathematics and engage with it as a sense-making activity. Second, when students get to secondary school they will not only be better prepared to engage with proving, which they will see as a natural extension of their prior mathematical experiences, but also better able to reason mathematically in disciplined ways, an important competence for meaningful engagement with mathematics more broadly.[1]

The Meaning of Proving

The idea of incorporating proving into students' mathematical experiences from the beginning of their schooling raises the issue of what could count as "proving" in school mathematics, especially at elementary school where proving has had a marginal place. This is an important issue: If we (researchers, teachers, curriculum developers, etc.) are unclear about what we mean by proving in (elementary) school mathematics, it is unrealistic to expect that we will be able to make good progress in teaching proving to students and in evaluating whether students have learned what we intended to teach them.

[1] An isomorphic argument has been offered for the importance of "early algebra" (e.g., Kaput, Carraher, & Blanton, 2007) in elementary school mathematics: The engagement of students with early algebraic ideas at elementary school may not only prepare them for secondary school algebra, which currently tends to be divorced from students' prior mathematical experiences, but may also enhance their learning of elementary school mathematics, notably elementary arithmetic (e.g., Carpenter et al., 2003; Carraher et al., 2006; Schoenfeld, 2007).

An Overview of the Situation in the Field

Despite the importance of proving in mathematics education research, there is a multiplicity of perspectives on and usages of the terms "proof" and "proving," especially with regard to the former. Some of these perspectives or usages have been summarized in Balacheff (2002), Stylianides et al. (2016b), and Weber (2008). Consider, for example, Balacheff's (2002) conclusion following a critical review of the relevant research literature:

> I went through a large number of research papers to figure out whether beyond the keywords we [researchers in mathematics education] had some common understanding [for the meaning of proof]. To discover that this is not the case was in fact not surprising. The issue then is to see where the differences are and what the price for them is in our research economy. My main concern is that if [we] do not clarify this point, it will be hardly possible to share results and hence to make any real progress in the field.
>
> Balacheff (2002, p. 1)[2]

Other mathematics education researchers have expressed similar concerns. Reid (2005), for example, noted that "if we [as a field] can acknowledge that there is an issue here, and discuss the characteristics of proof, we may be able to come to, if not agreement, then at least agreement on how we differ" (p. 465).

I think there should be no doubt that different research goals may be served better by different definitions of proof and proving, so it may be neither sensible nor desirable for all researchers to adopt common definitions. However, the important point seems to be that researchers should specify clearly the definitions of proof and proving that underpinned their research so as to allow proper interpretation of their findings and to facilitate comparisons across research reports. Consistent with this call for specificity, in what follows I offer and discuss the definitions of proof and proving that I use in this book.

These definitions derive from a conceptualization of the meanings of proof and proving that I proposed elsewhere (Stylianides, 2007b). While not comprehensive, the definitions are nevertheless well aligned with this book's remit. This is because the definitions were developed to be not only honest to mathematics as a discipline, but also (1) sufficiently "elastic," so as to allow description of proof and proving across all levels of education including the elementary school, and (2) pedagogically "sensitive," so as to support the study of the teacher's role in engaging students in proving (Stylianides, 2007b).

The Definition of Proving Used in this Book

According to the conceptualization in Stylianides (2007b), *proving* is defined broadly to denote the mathematical activity associated with the search for a proof. This definition of course raises the fundamental question of what counts as "proof," which I address below in the section "The Definition of Proof Used in this Book."

The activity of proving can involve a multiplicity of processes or auxiliary activities, which can support and give meaning to students' search for a proof (e.g., Alcock & Weber, 2010;

[2] The page number for this quotation refers to the online version of the paper.

Bartolini Bussi, 2000; Boero, Garuti, & Mariotti, 1996; Mason, 1982; Garuti, Boero, & Lemut, 1998; Lockwood, Ellis, Dogan, Williams, & Knuth, 2012; Pedemonte, 2007; Schoenfeld, 1985; Stylianides, 2008a; Weber & Alcock, 2004, 2009). Such processes or auxiliary activities include the following: engaging with inductive explorations to identify patterns or generalizations and make conjectures; working with particular cases or examples to test conjectures or gain a better understanding of what the conjectures mean and how they may be justified or refuted; using less formal ways of thinking (e.g., reasoning by analogy) or ways of representation (e.g., diagrams) to develop insight into arguments that may ultimately be developed into proofs; and using rhetorical means (not necessarily mathematical) to convince others about the epistemic value of a statement (this is often referred to as *argumentation*; see, e.g., Boero et al., 1996; Mariotti, 2006).

Zack's (1997) full report of what happened in her fifth-grade class during the implementation of the Squares Problem exemplifies virtually all of these processes or auxiliary activities, though the brief episode that I described in Chapter 1 also illustrates some of them. For example, in that episode we saw Will exploring particular cases to discover and test the criss-cross pattern, which offered a powerful visual way of representing the possible relationship between the side length of a square and the number of different squares of that size (Figure 1.2). This pattern allowed Gord to notice a pattern involving square numbers that served as the foundation for the conjecture that the total number of different squares in a square of a certain side length, say of size n, would be given by the sum of the first n square numbers. The episode illustrated further students engaging in argumentation, with Lew and Will, for example, employing different rhetorical means—both non-mathematical (making a bet) and mathematical (an argument showing that acceptance of a false claim leads to a contradiction)—to convince Rodd that his answer of 2310 for part 4 of the Squares Problem was wrong.

Students' engagement with proving can serve a broad range of functions (or purposes) including the following: justification or refutation (establishing the truth or falsity of an assertion); explanation (offering insight into why an assertion is true or false); discovery (inventing new results); communication (conveying results); systematization (organizing various results into a system of theorems); illustrating new methods of deduction; and defending the use of a definition or an axiom system (e.g., Bell, 1976; de Villiers, 1990, 1999; Larsen & Zandieh, 2008; Hanna & Barbreau, 2008; Stylianides, 2008a; Weber, 2002, 2010). The functions of justification/refutation and explanation have received particular attention in mathematics education research, not least because they are at the core of mathematics as a sense-making activity.

According to Harel and Sowder (2007, pp. 808–809), "[m]athematics as a sense-making activity means that one should not only ascertain oneself that the particular topic/procedure makes sense, but also that one should be able to convince others through explanation and justification of her or his conclusions," or indeed through refutation of others' (opposite) conclusions. With its justification/refutation function, proving can serve as a means for promoting conviction at both the individual and social levels by creating or removing doubts about the truth or falsity of an assertion (e.g., Harel & Sowder, 1998, 2007; Mason, 1982); with its explanation function, proving can serve as a means for promoting understanding by showing why an assertion is true or false (e.g., Ball & Bass, 2003; Hanna &

Jahnke, 1996). In the episode from Zack's class, Lew and Will engaged in a proving activity that served the functions of refutation and explanation: refutation, because the activity aimed to establish the falsity of Ross and Ted's claim following the challenge set before them, "*You have to prove us wrong*"; and explanation, because it aimed to help Ross and Ted understand why their claim did not make sense mathematically.

The Definition of Proof Used in this Book

According to the conceptualization in Stylianides (2007b), *proof* is defined in the context of a classroom community at a given time as follows:

> *Proof* is a *mathematical argument*, a connected sequence of assertions for or against a mathematical claim, with the following characteristics:
> 1. It uses statements accepted by the classroom community (*set of accepted statements*) that are true and available without further justification;
> 2. It employs forms of reasoning (*modes of argumentation*) that are valid and known to, or within the conceptual reach of, the classroom community; and
> 3. It is communicated with forms of expression (*modes of argument representation*) that are appropriate and known to, or within the conceptual reach of, the classroom community.
>
> Stylianides (2007b, p. 291) (emphasis in original)

This definition breaks down a mathematical argument into three *components*—the set of accepted statements, the modes of argumentation, and the modes of argument representation—and imposes on each of them certain requirements before the argument can be said to meet the standard of proof. Table 2.1 contains examples and non-examples of each of the three components of an argument.

In describing the requirements that the three components of an argument should satisfy for an argument to qualify as a proof, the definition seeks to achieve a balance between two important, albeit often competing, considerations: *mathematics as a discipline* and *students as mathematical learners* (Stylianides, 2007b,c). Regarding the consideration of mathematics as a discipline, the definition requires that proofs use true statements, valid modes of argumentation, and appropriate modes of representation, where the terms "true," "valid," and "appropriate" should be understood in the context of what is typically agreed upon in the field of mathematics (within the domain of particular mathematical theories). Regarding the consideration of students as mathematical learners, the definition requires that proofs depend on what is accepted by, or what is known or conceptually accessible to, a classroom community at a given time.

The merit of seeking to achieve a balance between these two considerations follows from my earlier discussion of the work of Bruner (1960) and Schwab (1978), as well as from the work of other general educational scholars such as Dewey (1902, 1903), all of whom emphasized the importance of the various subject areas at school being honest to their respective disciplines and honoring of students' current level. Support for this position is also found in the work of mathematics education researchers such as Mariotti (2006), who noted that "the crucial point that has emerged from different research contributions [in the

Table 2.1 Examples and non-examples of the three components of a mathematical argument mentioned in the definition of proof.

Components of an argument	Examples	Non-examples[a]
Set of accepted statements	• Definitions • Axioms or "local axioms"[b] • Assumptions • Theorems • Established procedures or rules (e.g., calculation methods)	• Conjectures
Modes of argumentation	• Correct application of valid logical rules of inference such as *modus ponens* and *modus tollens* (cf. *proof by contraposition*) • Correct use of definitions to derive a statement • Systematic consideration of all the cases (finitely many) involved in a situation (cf. *proof by exhaustion*) • Construction of an example that satisfies the conditions of a statement but violates its conclusion (cf. *proof by counterexample*) • Development of a reasoning that shows that acceptance of a statement leads to a contradiction (cf. *reductio ad absurdum* or *proof by contradiction*[c])	• Application of invalid rules of inference such as the inverse and converse "rules" • Acceptance of a statement based on the confirming but inconclusive evidence that is offered by the examination of some cases in the domain of the statement (cf. *empirical arguments*)
Modes of argument representation[d]	• Linguistic (e.g., verbal language) • Physical (e.g., concrete apparatus) • Diagrammatic/pictorial • Tabular (e.g., two-column format) • Symbolic (e.g., algebraic)	

[a] By a "non-example" I mean an instantiation of a component of an argument that would not normally be part of a proof.
[b] In school mathematics, students may take as axioms some statements that a mathematician would treat as theorems; these are often referred to as *local axioms* following Freudenthal's (1973) notion of "local organization." For example, a local axiom may be "the sum of the interior angles of a triangle (on the Euclidean plane) is 180°," which elementary students may use as a starting point to explore and prove different statements (theorems), such as statements about the interior angles of other polygons (Stylianides, 2007b). At a later stage when students develop their mathematical knowledge they can revisit a local axiom and try to prove it (by using, e.g., the same set of axioms as mathematicians), thus turning it into a theorem (de Villiers, 1986).
[c] By *reductio ad absurdum* I mean the method of proof that demonstrates a statement is *false* by showing that its acceptance leads to a contradiction. By *proof by contradiction* I mean the method of proof that demonstrates a statement is *true* by showing that acceptance of its *negation* leads to a contradiction.
[d] Depending on its use—appropriate versus inappropriate—each of these modes of argument representation can be an example or a non-example.

field of mathematics education] concerns the need for proof to be acceptable from a *mathematical* point of view but also to make sense for *students*" (Mariotti, 2006, p. 198; emphasis added). Furthermore, concrete images of what may be involved, or what it may look like, when mathematics teachers create classroom environments that aim to bridge the two considerations are found, for example, in the work of Ball (1993), Lampert (1990, 1992), and Zack (1997). A key part of the teacher's role in creating these environments is to "consider the mathematics in relation to the children and the children in relation to the mathematics," with the teacher's ears and eyes "search[ing] the world around us, the discipline of mathematics, and the world of the child with both mathematical and child filters" (Ball, 1993, p. 394).

I now turn back to the episode from Zack's class that I described in Chapter 1 to consider whether the argument that Lew and Will offered against Ross and Ted's claim could be considered to be a proof. I remind the reader that Ross and Ted claimed that the total number of different squares in a 60-by-60 square is 2310, which they obtained by taking the total number of different squares in a 10-by-10 square (i.e., 385) and multiplying that number by 6. Lew and Will's argument went as follows: Even if one considered only the "little squares" (i.e., the 1-by-1 squares) in a 60-by-60 square, their number would be $60 \times 60 = 3600$, which is already bigger than 2310.

According to the definition of proof, this argument qualified as a proof against the particular claim in the given classroom context because of the following:

(1) It used true statements that were readily accepted by the classroom community, notably the fact that there are $60 \times 60 = 3600$ "little squares" in a 60-by-60 square (the easily inferred justification for the product is that there are 60 rows with 60 little squares in each row).

(2) It employed a valid mode of argumentation that was understandable to the classroom community and was associated with the development of a reasoning that shows that acceptance of a claim leads to a contradiction (the total number of squares in a 60-by-60 square cannot be 2310, as the total number is definitely more than 3600).

(3) It was represented using appropriate oral language that students in the class could follow.

The above proof aimed to refute a false claim. In the episode we can infer another argument that aimed to justify a true claim and could also count as a proof. This argument relates to the work of Lew and Ross in part 1 of the Squares Problem. The two students used variants of a "trace and count" approach, with Lew in particular sliding each size square across the master grid and counting the number of squares of that size. Although Zack (1997) does not give many details about Lew's work in her report, Lew's approach seemed to have involved systematic sliding and counting of the squares of every possible size in a 4-by-4 square, yielding the following results: in a 4-by-4 square there are sixteen 1-by-1 squares; nine 2-by-2 squares; four 3-by-3 squares; and one 4-by-4 square. According to the definition, an argument based on this approach would qualify as a proof for the claim that there are a total of 30 different squares in a 4-by-4 square, because of the following:

(1) It used true statements that were readily accepted by the classroom community, such as that each new square of a certain size corresponded to a new position of the "sliding square" in Lew's approach.
(2) It employed the valid mode of argumentation associated with systematic consideration of all the cases (finitely many) involved in a situation (the systematic sliding and counting in Lew's approach helped ensure that all of the different squares were considered and that none of them was counted more than once).
(3) It was represented using appropriate oral language and a physical demonstration (sliding square) that students in the class could follow.

This proof for the 4-by-4 square can easily be extended for a square of any size, provided that the action of "sliding" is done in one's imagination rather than physically. In a project I conducted with 14–15-year-olds in England, the students worked on a variant of the Squares Problem (Stylianides, 2009a; Stylianides & Stylianides, 2014b) and used the idea of a sliding square in their proofs about the number of different 3-by-3 squares in a 60-by-60 square. In particular, they considered a 3-by-3 square on the upper left side of a 60-by-60 square, as presented in Figure 2.1, and they imagined sliding it over to the right, one place at a time, until it reached the upper right side of the 60-by-60 square. Given that there are 57 $(=60-3)$ available positions for it to slide over, there are another 57 3-by-3 squares in the "top row" of the 60-by-60 square, i.e., a total of 58 $(=57+1)$ 3-by-3 squares. The students also found how many such rows there are by imagining the same 3-by-3 square sliding down

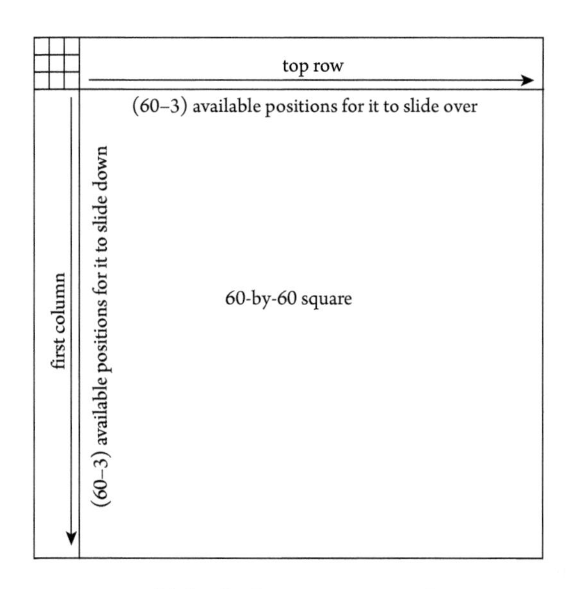

Figure 2.1 A "sliding" 3-by-3 square in a 60-by-60 square.

the "first column" of the 60-by-60 square. The answer was again 58. They concluded that there are 58 rows with 58 3-by-3 squares in each row, which makes a total of 58^2 ($= 58 \times 58$) 3-by-3 squares in a 60-by-60 square.

Exactly the same argument could apply for the number of 3-by-3 squares in a larger square of any size, say N-by-N, so long as one understood that the operations on the 60-by-60 square that were described previously were not specific to it having this particular size. A general argument about the number of different 3-by-3 squares in a larger size square that would be formulated using the particular case of the 60-by-60 square as a representative of the whole class of N-by-N squares would be an example of a "generic argument" (e.g., Balacheff, 1988a; Mason & Pimm, 1984; Movshovitz-Hadar, 1988; Rowland, 2002; Tall, 1999). In more general terms, *generic arguments* are defined as arguments that justify the truth of a statement about a set of cases by showing that the statement holds for a particular case (or example) in the domain of the statement that possesses no special properties so that the reasoning used for that case can be applied to any other case.

Here is another example of a generic argument that was offered by an elementary student called Jamie for the statement "the sum of any two odd numbers is an even number":

> See, here is the odd number. [*She makes a collection of pairs of blocks together with one additional block.*] And here is another. [*She makes another collection of pairs of blocks together with one additional block. In both cases she does not bother to count the pairs of blocks, so she does not know how many blocks she has. She only knows that they represent an odd number because there is a block left over.*] Now we put them together, and we still have all these pairs, but now these two pair up with each other. [*She puts the two single blocks together to make another pair.*] See, all the blocks are paired up. So the number is even.
>
> Carpenter et al. (2003, p. 90) (emphasis in original)

Generic arguments such as this one use a valid mode of argumentation, as they capture adequately the generality of the statement they aim to justify, and normally satisfy the definition of proof subject to the other standard requirements, notably the use of mathematical resources (e.g., relevant definitions) that are known or accessible to the students in a classroom community at a given time (Stylianides, 2007b). Many researchers view generic arguments as a means of helping students see the general in the particular, and as a bridge toward more general arguments that are independent of particular cases (e.g., Leron & Zaslavsky, 2013; Malek & Movshovitz-Hadar, 2011; Mason & Pimm, 1984; Rowland, 2002; Stylianides, 2009b).

I now move on to present a non-example of a proof, using again the context of the Squares Problem from the episode in Zack's class. This non-example is an extension of Will's work: Will discovered the criss-cross pattern about the number of different squares of each size in a 4-by-4 square (Figure 1.2); he tested it on the 5-by-5 square and found that it worked again; and he subsequently acted in accordance with the assumption that the pattern would work in the same way for squares of any size. Thus one would expect Will to argue, for example, the following: the number of different squares of each size in, say, a 7-by-7 square would be as presented in Figure 2.2, because the criss-cross pattern would apply in this case as it did in all of the other cases examined previously.

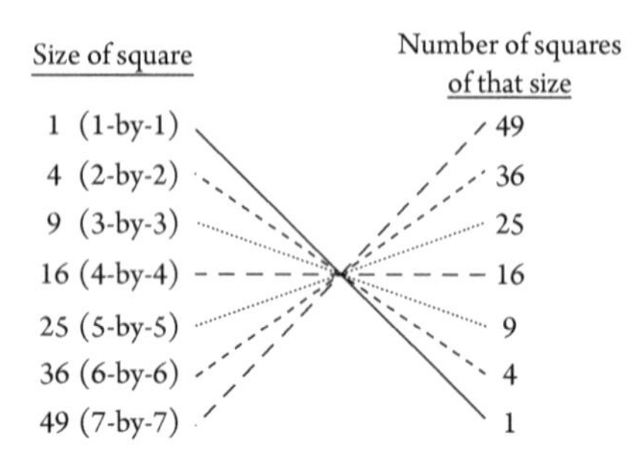

Figure 2.2 Will's criss-cross pattern applied in the case of a 7-by-7 square.

Such an argument would *not* qualify as a proof because it uses an invalid mode of argumentation. Specifically, the argument is based on the mathematically problematic reasoning that a pattern that emerged from the examination of a few cases can be accepted as true and applied securely in a new case. Arguments that are based on this reasoning belong to the category of arguments often referred to as "empirical" (e.g., Balacheff, 1988a; Harel & Sowder, 1998). More precisely, *empirical arguments* are defined to be those arguments that purport to show the truth of a statement based on the confirming yet inconclusive evidence that is offered by the examination of some cases in the domain of the statement.[3] Unlike a generic argument, which uses a particular case as a context to formulate a justification that can readily be applied for all other cases in the domain of a statement, an empirical argument simply confirms a statement in some cases and purports, on this weak basis, that the statement holds for all cases in its domain. Clearly, an empirical argument requires one to make a leap of faith.

The definition of proof excludes empirical arguments from being considered as proofs at any level of education due to their use of invalid modes of argumentation. The exclusion of empirical arguments from the class of proofs should not be interpreted as a lack of appreciation of the important role that examples play in the proving activity. Indeed, the generation or study of examples (including counterexamples) is instrumental in revealing or illuminating underlying mathematical structures, in exploring or defining boundaries of generalizations, and in making or refining conjectures (e.g., Lannin, Ellis, & Elliott, 2011; Lin & Tsai, 2012; Reid, 2002; Watson & Chick, 2011; Zaslavsky, 2014; Zazkis, Liljedahl, & Chernoff, 2008). One might wonder, then, whether by excluding empirical arguments from being considered as proofs at any level of education, the definition sets the standard of proof too high, especially for elementary students. The mounting body of research showing that many

[3] In the literature, empirical arguments are typically used to describe arguments for statements formulated as generalizations over *infinite* sets. In these situations, the cases examined in an empirical argument are a proper subset of all the possible cases in the domain of the generalization. Yet, the term "empirical" can also apply to arguments for statements formulated over *finite* sets. In these situations, the cases examined in an empirical argument can again be a proper subset of all the possible cases, but they can also be the full set of possible cases. If the latter happens, the empirical argument would fail to justify logically that all the possible cases have indeed been examined, thus still offering "confirming yet inconclusive evidence" for the truth of the statement.

secondary and university students consider empirical arguments as proofs might exacerbate this concern (e.g., Buchbinder & Zaslavsky, 2007; Coe & Ruthven, 1994; Chazan, 1993; Edwards, 1999; Goetting, 1995; Goulding, Rowland, & Barber, 2002; Goulding & Suggate, 2001; Healy & Hoyles, 2000; Knuth, Choppin, & Bieda, 2009; Knuth et al., 2002; Küchemann & Hoyles, 2001–2003; Martin & Harel, 1989; Moore, 1994; Morris, 2002; Sowder & Harel, 2003). However, psychological research on the cognitive development of young children's capacity for deductive reasoning (reviewed in Stylianides & Stylianides, 2008b), together with mathematics education research on the teaching and learning of proving in supportive classroom environments at elementary school (e.g., Ball & Bass, 2000b, 2003; Lampert, 1990, 1992; Maher & Martino, 1996; Reid, 2002; Stylianides, 2007a–c; Zack, 1997), all suggest collectively that non-empirical modes of argumentation *are* within the conceptual reach of elementary students and that instruction plays a crucial role in enabling that to happen. Indeed, students' difficulties in understanding that empirical arguments are not proofs may be attributed less to cognitive-developmental constraints and more to the limited opportunities that typical classroom instruction offers to students to develop their understanding of proof.

A consistent meaning of proof as a non-empirical argument throughout school mathematics, such as the one promoted by the definition, not only honors mathematics as a discipline wherein empirical arguments are not proofs, but also sets the basis for the development of a conception of proof at elementary school that would not have to be unlearned at secondary school. According to Martin and Harel (1989), if elementary "teachers lead their students to believe that a few well-chosen examples constitute a proof, it is natural to expect that the idea of proof in high school geometry and other courses will be difficult for students" (pp. 41–42). The general idea of trying not to instill in students faulty or unproductive conceptions during their introduction to a subject finds support in the work of Dewey (1903). According to him, whatever the preliminary approach to learning is, it should not inculcate "mental habits and preconceptions which have later on to be bodily displaced or rooted up in order to secure proper comprehension of the subject" (Dewey, 1903, p. 217).

To conclude, by proposing a meaning of proof that preserves some fundamental mathematical expectations irrespective of educational level (such as a proof being a non-empirical argument), the definition supports continuity and coherence throughout students' schooling in their mathematical experiences with proof. At the same time, however, the definition allows for what may count as a proof to be partly determined by the particular circumstances of a classroom community at a given time. By so doing, the definition also supports a rather elastic meaning of proof that can evolve in parallel with the developing mathematical knowledge of students over their school years.

The Place of Proving in Mathematics Classrooms

In this section I first discuss the current place of proving in secondary and elementary mathematics classrooms internationally, also reflecting on possible reasons for which this place is marginal. I then argue that mathematics tasks can play a significant role in efforts to elevate the place of proving in students' mathematical work at elementary school and beyond.

The Current Place of Proving

A number of large-scale classroom-based studies, such as the 1995 and 1999 Video Studies in the series Trends in International Mathematics and Science Study (TIMSS), showed that proving has a marginal place in typical secondary mathematics classrooms around the world. Consider, for example, the findings of Hiebert et al. (2003), who analyzed the kinds of reasoning encouraged by the mathematics tasks used in the lessons of the TIMSS 1999 Video Study, including those tasks that involved proofs. A task was considered to involve a proof "if the teacher or students verified or demonstrated that the result must be true by reasoning from the given conditions to the result using a logically connected sequence of steps" (Hiebert et al., 2003, p. 73). They found that such tasks were evident to a substantial degree only in Japanese lessons, in which on average 26% of the tasks per lesson involved proofs and 39% of the lessons included at least one proof. In the Czech Republic, Hong Kong SAR, and Switzerland the corresponding percentages were considerably lower (1, 2, and 3% and 5, 12, and 11%, respectively), while in Australia, the Netherlands, and the United States there were too few instances of proofs to allow the calculation of reliable estimates.

Smaller-scale classroom-based studies that were conducted more recently in individual countries, such as those by Bieda (2010) and Sears (2012) in the United States, revealed a similarly inadequate treatment of proving in secondary classrooms, even in classrooms where the conditions for students to engage with proving were quite favorable. This state of affairs in secondary mathematics classrooms helps explain, at least partly, the disappointing findings of large-scale surveys of secondary students' understanding of proof (e.g., Knuth et al., 2009; Küchemann & Hoyles, 2001–2003; Senk, 1989). For example, the longitudinal survey conducted by Küchemann and Hoyles (2001–2003) involving a national sample of 1512 high-attaining secondary students in England showed only modest (if any) improvements over time in students' understanding of proof. Senk (1989) reported similarly disappointing findings from a survey of 1520 US secondary students taking a geometry class: Only 30% of the students were able to prove at least three statements out of the four given to them, despite the fact that two of the expected proofs required only a single deduction beyond the hypotheses; 29% of the students were unable to construct a single proof.

To the best of my knowledge, there are no large-scale studies documenting the place of proving in typical elementary mathematics classrooms. Yet the bleak picture that research has painted about the marginal place of proving in secondary mathematics classrooms around the world allows little (if any) doubt that the situation in elementary classrooms is similar if not worse. Indeed, it appears to be common knowledge among practitioners and researchers internationally (e.g., Ball et al., 2002) that elementary students in typical classrooms have limited opportunities to learn about proof, while policy makers and curriculum frameworks in different countries (e.g., Department for Education, 2013; NCTM, 2000) are calling for an elevated place of proving in the mathematical experiences of all elementary students. As I explained previously, an elevated place of proving in elementary mathematics classrooms could not only enhance elementary students' mathematical learning by allowing them better access to mathematics as a sense-making activity, but could also reduce the gap between their typical and desirable understandings of proof by equipping them with a stronger foundation for proving at their entry to secondary school.

Before I consider what it might take to elevate the place of proving in elementary students' mathematical work, I reflect on factors that might have contributed to the marginal place of proving at elementary school. While I recognize that this place has probably resulted from a synergy of many factors, I have singled out for brief discussion four factors that appear to be particularly important. These factors relate to the following: the weak knowledge that many elementary teachers have about proof (factor 1) and their presumed beliefs that proving is an advanced mathematical topic beyond the reach of elementary students (factor 2); the high pedagogical demands placed on elementary teachers who strive to engage their students in proving (factor 3); and the inadequate instructional support offered or available to elementary teachers about how to achieve that goal in their classrooms (factor 4). Factors 1 and 2 are not presented to support a deficit model of teacher competence, and indeed such an interpretation would be unfortunate. Rather, I present these factors to emphasize, alongside factor 3, the challenges involved for non-specialist teachers of mathematics (as most elementary teachers are) when charged with the demanding endeavor of engaging young children in proving, a hard-to-teach and hard-to-learn mathematical activity. Accordingly, the locus of responsibility shifts from factors 1–3, which concern mostly elementary teachers, to factor 4, which concerns several stakeholders, such as curriculum developers (including textbook authors) and teacher educators, but also researchers who could aim to provide the relevant research basis for the required instructional support.

Factor 1: teachers' knowledge

The first factor relates to the mathematical knowledge of elementary teachers about proof. While a rather broad and deep knowledge about different aspects of proof is generally required for teachers to be able to successfully engage their students in proving (Stylianides & Ball, 2008), a number of studies have shown that elementary teachers (pre-service or in-service) tend to have weak knowledge about proof. A major issue is the difficulty that teachers have in distinguishing between valid and invalid modes of argumentation. For example, elementary teachers were found to consider empirical arguments as proofs of mathematical generalizations (e.g., Goetting, 1995; Goulding et al., 2002; Martin & Harel, 1989; Morris, 2002, 2007), to be unconvinced by the power of a single counterexample to refute a false generalization (e.g., Simon & Blume, 1996; Stylianides, Stylianides, & Philippou, 2002), and to consider that a conditional statement is equivalent to its inverse (e.g., Goetting, 1995; Stylianides, Stylianides, & Philippou, 2004) but not to its contrapositive (Stylianides et al., 2004).

As I mentioned earlier, these findings should not be interpreted as a criticism of elementary teachers as individuals or professionals. After all, elementary teachers are not mathematics specialists. Also, similar difficulties in distinguishing between valid and invalid modes of argumentation were reported, albeit to a lesser extent, by studies with secondary mathematics teachers (pre-service or in-service) and undergraduate mathematics majors (e.g., Knuth, 2002a; Ko & Knuth, 2009, 2013; Sowder & Harel, 2003; Stylianides et al., 2004, 2007).

Factor 2: teachers' beliefs

The second factor relates to the beliefs of elementary teachers about the role of proof and proving in elementary school mathematics. While I am not aware of any studies that have specifically examined this aspect among elementary teachers, I think it is quite safe to extrapolate from the findings of relevant studies with secondary mathematics teachers and mathematics professors. Secondary mathematics teachers were found to view proof in a pedagogically limited way and tended to consider proving as an appropriate goal for only a few students who are "developmentally ready" for it (e.g., Bieda, 2010; Iscimen, 2011; Knuth, 2002b). Similar to secondary mathematics teachers, some mathematics professors questioned whether proving was an appropriate goal for all mathematics majors, as they felt that some of these undergraduate students were not able to understand proof (e.g., Alcock, 2010; Harel & Sowder, 2009; Weber, 2012). In light of these findings one can reasonably expect many elementary teachers, who teach younger students and who themselves tend to face difficulties with proof, to view proving as an advanced mathematical topic that is beyond the reach of elementary students.

Of course, whether or not proving is, or can be, an appropriate goal for students at any level of education depends on what one takes proof and proving to mean, especially proof. I argue that a meaning of proof like the one I discussed earlier, which views proof as a vehicle to mathematical sense-making and as an argument that is not only honest to the discipline of mathematics but also sufficiently elastic to adapt to students' current level or knowledge (Stylianides, 2007b), has a better chance of being accepted as appropriate for elementary or older students than a narrower meaning that would equate proof with "formal deductions," for example.

Factor 3: pedagogical demands

The third factor relates to the high pedagogical demands placed on teachers who strive to create in their classrooms meaningful learning opportunities for their students to engage in proving. Research showed that creating and effectively managing these learning opportunities for students might be challenging and complicated by dilemmas and uncertainties, even for experienced elementary teachers (e.g., Ball, 1993; Heaton, 2000; Lampert, 2001; Zack, 1997).

Research showed further that, even if the knowledge and beliefs of pre-service elementary teachers were attuned to teaching proof, they would still encounter major obstacles as they embarked on pedagogical practices that put a premium on proving (Stylianides, Stylianides, & Shilling-Traina, 2013). These obstacles relate to a number of challenges, including the following that are well documented in the broader literature, especially in studies that have examined the use of proving tasks or other kinds of mathematically rich, cognitively demanding tasks: facilitating student work on these tasks without lowering the tasks' cognitive demands (e.g., Boston & Smith, 2009; Davis & McKnight, 1976; Doyle, 1988; Henningsen & Stein, 1997; Sears & Chávez, 2014; Stein, Grover, & Henningsen, 1996; Stylianides & Stylianides, 2008a); managing classroom dialogue and bringing together different student contributions during whole-class discussions (e.g., Alexander, 2006; Kazemi, Franke, & Lampert, 2009; Lampert, 2001; Stein, Engle, Smith, & Hughes, 2008; Stylianides

& Stylianides, 2014b); and in-the-moment decision making about how to respond to student contributions or other kinds of classroom events while also steering classwork toward the intended learning goals (e.g., Rowland, Hodgen, & Solomon, 2015; Scherrer & Stein, 2013; Stein et al., 2008; Stylianides & Stylianides, 2014b; Sullivan & Mousley, 2001; van Es & Sherin, 2002).

Factor 4: instructional support

The fourth factor relates to the inadequate instructional support offered or available to elementary teachers about how to productively engage their students in proving. This instructional support could take, for example, the form of well-designed textbooks (or other curricular resources) that would offer to teachers suggestions for mathematics tasks they could use to engage their students in powerful mathematical activity (including proving), as well as guidance about how they could implement these tasks and how they could effectively manage their students' activity (e.g., Ball & Cohen, 1996; Davis & Krajcik, 2005; Davis, Palincsar, & Arias, 2014; Stylianides, 2008b, 2014, 2016). The reality of the situation, though, is that there is a scarcity of well-designed textbooks offering effective support to teachers on how to engage their students in proving. This applies across school levels, elementary (Bieda, Ji, Drwencke, & Picard, 2014) and secondary (Davis, Smith, & Roy, 2014; Fujita & Jones, 2014; Otten, Males, & Gilbertson, 2014; Stylianides, 2008b, 2009b; Thompson, Senk, & Johnson, 2012). For example, Bieda et al. (2014) found that, on average, only 3.7% of the mathematics tasks in seven upper elementary mathematics textbooks in the United States included opportunities for students to engage in proving. This state of affairs is problematic given the important role that textbooks play, or can play, in the everyday practice of many teachers and in the learning opportunities that teachers offer to students to learn mathematics in general and proving in particular (e.g., Bieda, 2010; Cai, Ni, & Lester, 2011; Mullis, Martin, Foy, & Arora, 2012; Stylianides, 2014, 2016; Tarr, Chávez, Reys, & Reys, 2006). Indeed, according to TIMSS 2011 (Mullis et al., 2012), textbooks are the most frequent basis of mathematics instruction at both fourth and eighth grades, being used on average with approximately 75% of students world-wide.

Elementary teachers appear to also receive limited instructional support from curriculum frameworks, including some that place a premium on proving as a learning goal for all students. Take, for example, the recently published English national curriculum for mathematics (Department for Education, 2013), which, as I explained earlier, set a core aim specifically related to proving for students of all ages: "[All students should] reason mathematically by following a line of enquiry, conjecturing relationships and generalisations, and developing an argument, justification or proof using mathematical language" (Department for Education, 2013, p. 3). While this is one out of only three core aims set by the curriculum for students of all ages, the curriculum offered no specific guidance to teachers about how they could promote this aim in their classrooms. In particular, and contrary to how the curriculum treated "standard topics" in elementary mathematics, such as number and operations, the curriculum did not break down the aim into more specific, age-appropriate learning goals that teachers could target in their everyday practice.

It is unclear what kind of instructional support is received by elementary teachers in mathematics teacher education (either during their training or afterward), though there is weak evidence to suggest that there may not be much of it. In particular, in an analysis of 16 university textbooks designed specifically for use in mathematics courses for elementary teachers in the United States, we found the treatment of issues related to proving to be only loosely connected with the work of teaching in elementary classrooms (McCrory & Stylianides, 2014). To the best of my knowledge, there is no research on the actual use of those textbooks in teacher education or on what happens in pedagogical courses in the United States or elsewhere.

Of course, and to be fair to textbook authors, curriculum developers, and teacher educators, the apparent or presumed scarcity of instructional support for elementary teachers to teach proving mirrors to a large extent the limited research knowledge currently available about how to productively engage elementary students in proving (Stylianides et al., 2016a, b). On a positive note, the role and place of proving at the secondary school and university levels has received more attention, with some research findings casting light on what it would take to design theory-based interventions in classroom settings to promote understanding of proof by secondary or university students (Anderson, Corbett, Koedinger, & Pelletier, 1995; Brown, 2014; Harel, 2002; Hodds, Alcock, & Inglis, 2014; Jahnke & Wambach, 2013; Mariotti, 2000, 2013; Schoenfeld, 1985; Stylianides & Stylianides, 2009b, 2014b; Weber, 2006).

The Role of Mathematics Tasks in Elevating the Place of Proving

The interdependence and multiplicity of factors that have contributed to the marginal place of proving in elementary mathematics classrooms imply that there are no easy solutions to the problem of elevating the status of proving in elementary students' mathematical work. Notwithstanding the complexity of the problem, mathematics tasks offer a point of leverage to help address it, because such tasks have a major impact on the work that takes place in classrooms at elementary school and beyond (e.g., Boston & Smith, 2009; Christiansen & Walther, 1986; Doyle, 1988; Leinhardt, Zaslavsky, & Stein, 1990; Mason, Watson, & Zaslavsky, 2007; Prusak, Hershkowitz, & Schwarz, 2012; Sears & Chávez, 2014; Sullivan, Clarke, & Clarke, 2013; Zaslavsky, 2005). In particular, the mathematics tasks used in a classroom can limit or broaden students' views of the subject matter they engage with (e.g., Henningsen & Stein, 1997; Schoenfeld, 1992), and, together with the actions performed by the teacher during their implementation, they "constitute the major method by which mathematics is expected to be conveyed to the students" (Christiansen & Walther, 1986, p. 244).

Thus, unless elementary teachers have access to, or design on their own, proving tasks to use in their classrooms, it would be unrealistic to expect that elementary students would be offered productive opportunities to engage in proving. However, as I explained earlier, there is scarcity of proving tasks in elementary mathematics textbooks (Bieda et al., 2014). Also, the weaknesses in many elementary teachers' mathematical knowledge about proof and the lack of a pedagogically functional meaning of proving in elementary school mathematics, which gives ground to views of proof as an advanced topic beyond the reach of elementary students, make it unlikely that elementary teachers would design independently an appropriate collection of proving tasks to use in their classrooms. Further, that would be an unfair expectation of elementary teachers, especially in the midst of their busy professional lives, for these teachers would have had to navigate a relatively unexplored curricular territory about which the available research basis is also limited. In this book I take a step toward addressing this gap in the body of the research.

In the following chapters I propose and study a categorization of proving tasks that elementary teachers can use in their classrooms in order to create productive opportunities for their students to engage in proving. In particular, I explore the interplay between different kinds of proving tasks and the proving activity that tasks of each kind might help generate in the classroom, while also considering the role of teachers in implementing the tasks and in supporting their students' work. Given the issues that I discussed earlier in relation to the four factors that might have contributed to the marginal place of proving in elementary classrooms today, the work reported in the book can have important implications for elevating that place. I discuss these implications in Chapter 8.

3

The Set-up of the Investigation

In this chapter I set up the investigation that I undertake in subsequent chapters. Specifically, I describe a categorization of proving tasks and I identify main characteristics of these tasks that can influence proving activity during their implementation in the classroom.[1] I also describe the data sources and analytic method used in the book. Some of the issues discussed in this chapter can be relevant throughout school mathematics, though most of the examples I offer relate to the elementary school level.

A Categorization of Proving Tasks

A major finding in the literature on whole-number arithmetic has, I believe, broader implications for the design of mathematics tasks. According to this literature, elementary students should be offered opportunities to engage in different kinds of arithmetic tasks (often presented as "word problems") so that they can develop a broad-based and deep understanding of the different meanings of the four basic operations (e.g., Carpenter, Fennema, & Franke, 1996; Fischbein, Deri, Nello, & Marino, 1985; Haylock, 2014). Indeed, the overemphasis of typical classroom instruction on certain kinds of arithmetic tasks that promote certain meanings of the basic operations creates a cognitive obstacle to students learning other meanings. For example, it was found that when students' experiences with division were dominated by tasks that promoted the "partitive" meaning of division they had difficulties with tasks that promoted its "measurement" meaning (Fischbein et al., 1985).

Making an analogy to the area of proving, I argue that a broad-based and deep understanding of proving requires that students engage in different kinds of proving tasks so that they benefit from the complementary nature and learning affordances of those tasks. The issue that is raised, then, is what might be a possible categorization of proving tasks that can serve this purpose?

[1] In discussing these issues I use with permission parts of Stylianides and Ball (2008) (license number 3454130366327).

One could classify proving tasks according to a range of criteria, such as the following: the potential of a task to create a situation of conflict or uncertainty for students (e.g., Hadas, Hershkowitz, & Schwarz, 2000; Prusak et al., 2012; Stylianides & Stylianides, 2009b, 2014a; Zaslavsky, 2005); the extent to which a task places proving activity in a broader conjecturing process that gives rise to the statement which is then subject to proof (e.g., Bartolini Bussi, 2000; Boero et al., 1996; Garuti et al., 1998; Mason, 1982; Stylianides, 2008a); the extent to which a task allows multiple solutions that can draw on various mathematical tools, possibly from different mathematical domains (e.g., Kondratieva, 2011; Leikin, 2010; Leikin & Levav-Waynberg, 2008); or the extent to which a task places proving activity in the context of diagrams whose transformations can allow, for example, the generation of counterexamples (e.g., Komatsu & Tsujiyama, 2013; Komatsu, Tsujiyama, Sakamaki, & Koike, 2014). These are all important criteria for categorizing proving tasks, not least because of the pedagogical issues they draw attention to when organizing or studying proving activity in the classroom. Yet the criteria were developed in the service of particular research goals, and so the possible use of one or more of them in the categorization of proving tasks for the book could compete with my interest in a rather encompassing perspective on proving tasks in the elementary classroom.

I decided to categorize proving tasks according to the following two *mathematical* criteria. First, the number of cases involved in a task: a single case, multiple but finitely many cases, or infinitely many cases. Second, the specific purpose (or function) of proving served in the task: to justify or to refute a statement. These criteria can apply to virtually any proving task, thus supporting, as I had intended, a rather encompassing perspective on proving tasks in the elementary classroom. Indeed, the selected criteria do not exclude consideration of the sort of pedagogical issues highlighted by the criteria I summarized earlier; rather, they allow space for these issues to come into play if and when appropriate. Also, as I shall explain in the section "Characteristics of Proving Tasks That Can Influence Their Generated Proving Activity," the criteria offer a useful lens for examining the proving activity that different kinds of proving tasks can potentially generate in the classroom.

Table 3.1 offers illustrative examples of proving tasks that satisfy the two criteria in different ways. I have chosen examples from the same mathematical territory (which relates to the concepts of prime numbers and multiples of a number) in order to highlight differences between categories. More examples of tasks are presented and discussed later in this chapter and in subsequent chapters.

Next I discuss four points in regard to the proposed categorization. First, even though a classification criterion is the number of cases involved in a proving task, this does not mean that the focus of any proving task is, or should be, on the cardinality of the set of cases involved in it. For example, cardinality is not a focus in task (i) in Table 3.1; rather, the focus there is on a property of the elements of the (infinite) set $\{m|m = 3k + 3l, k, l, \in \mathbf{Z}\}$. I clarify that, in the context of the proposed categorization, the number of cases involved in a task refers to the cases covered in the relevant statement that is mentioned in the task—as opposed, for example, to the number of cases that one might have to consider in constructing a proof for that statement. Take, for example, task (b), which calls for justification that 73 is

Table 3.1 A classification of proving tasks with illustrative examples.

Number of cases involved in a proving task	Purpose of a proving task	
	Justification of a statement	**Refutation of a statement**
A single case	(a) Prove that 33 plus 36 is a multiple of 3 (b) Prove that 73 is a prime number	(c) Disprove that 33 plus 36 is a multiple of 6 (d) Disprove that 51 is a prime number
Multiple but finitely many cases	(e) Prove that the sum of any two multiples of 3 between 30 and 50 is a multiple of 3 (f) Prove that there are four prime numbers between 10 and 20	(g) Disprove that the sum of any two multiples of 3 between 30 and 50 is a multiple of 6 (h) Disprove that there are as many prime numbers between 1 and 10 as there are between 21 and 30
Infinitely many cases	(i) Prove that the sum of any two multiples of 3 is a multiple of 3 (j) Prove that six times any prime number is a multiple of 3	(k) Disprove that the sum of any two multiples of 3 is a multiple of 6 (l) Disprove that the sum of any two prime numbers is a prime number

a prime number. The statement involves only the number 73 (i.e., a single case), but a proof for this statement would normally require one to examine whether or not every element in a specific set of positive integers is a factor of 73 (i.e., the examination would be likely to consider multiple cases).[2]

Second, the purposes of justification and refutation that are considered in the categorization of proving tasks are, as discussed in Chapter 2, two out of many important purposes (or functions) that proving can serve in the classroom, including explanation, discovery, communication, and systematization (e.g., Bell, 1976; de Villiers, 1990, 1999). I see these other purposes as extending over the two primary purposes of justification and refutation. Indeed, a proving activity that aims to justify or refute a mathematical statement can also offer insight into why the statement is true or false (explanation purpose) as well as help invent or convey new results (discovery and communication purposes, respectively). Also, work on a collection of related proving tasks, such as tasks (a), (e), and (i) in Table 3.1, can help promote the purpose of systematization through organization of related mathematical results in increasing order of mathematical generality.

[2] The set of positive integers to be examined can depend on students' prior knowledge. For example, if students understand that a composite number has at least one factor greater than 1 and lower than or equal to its square root, then it would suffice for students working on task (b) to examine all positive integers from 2 to 8 (inclusive).

Third, the number of cases involved in a proving task as well as the purpose of the task may not be transparent, or made explicit, to solvers. For example, tasks (i) and (j) in Table 3.1 do not explicitly specify to solvers the number of cases involved in them, though solvers can infer that each of these tasks involves an infinite set provided that they know there are infinitely many pairs of multiples of 3 and infinitely many prime numbers, respectively. Task (f) appears to specify to solvers the set of cases involved in it (i.e., all "numbers" between 10 and 20), but even in this task, unless solvers know that a prime number can only be a positive integer, they may be unclear about the number of cases involved in the task. As far as the purpose of a task is concerned, while for the sake of illustration I have phrased all the tasks in Table 3.1 in a way that makes explicit their specific purpose (i.e., to justify or to refute a statement), this purpose could have been concealed from solvers. For example, the tasks could start with, "Prove or disprove that. . . ."

Fourth, and closely related to the previous point, when students engage in proving in the broader context of mathematics as a sense-making activity, they are not normally given ready proving tasks in the form presented in Table 3.1: "Prove that. . ." or "Disprove that. . . ." Rather, proving tasks tend to arise naturally from students' own work as they explore mathematical phenomena, examine particular cases, discuss alternative ideas, and generate conjectures (e.g., Balacheff, 1990; Hadas et al., 2000; Lampert, 1990; Lin & Tsai, 2012; Mason, 1982; Mogetta, Olivero, & Jones, 1999; Schoenfeld, 1985; Stylianides, 2008a; Watson & Mason, 2005). The Stamps Problem in Table 3.2 describes a situation that has the potential

Table 3.2 The Stamps Problem, as well as conjectures and corresponding proving tasks likely to arise naturally from students' work on the problem.

The problem	You have lots of 1¢, 2¢, and 3¢ stamps (¢ is the symbol for cent, a monetary unit). You can use the available stamps to make different amounts. For example, you can make 3¢ in three different ways: (1) by using only a 3¢ stamp; (2) by using one 1¢ stamp and one 2¢ stamp (the order doesn't matter); and (3) by using three 1¢ stamps. These are all the different ways there are to make 3¢. Explore how many different ways there are to make different amounts using the available stamps. Prove your answers.
Possible conjectures and corresponding proving tasks	*Conjecture 1*: "You can make 4¢ in four different ways." (true) • The corresponding proving task can engage students in *justification* of a statement that involves *multiple but finitely many cases* *Conjecture 2*: "You can make 5¢ in five different ways." (true) • The corresponding proving task is the same as for Conjecture 1 *Conjecture 3*: "You can make 6¢ in six different ways." (false) • The corresponding proving task can engage students in *refutation* of a statement that involves *multiple but finitely many cases* *Conjecture 4*: "You can make *N*¢ in *N* different ways (where *N* is any natural number)." (false) • The corresponding proving task can engage students in *refutation* of a statement that involves *infinitely many cases*

to naturally engage students in proving activity, in the context of different conjectures corresponding to different kinds of proving tasks.

To further illustrate the fourth point, I consider the following episode from Lampert's (1990) fifth-grade class in the United States. The episode presents a situation where students' exploration of a single case led naturally not only to a proving task concerning that particular case, but also to a proving task concerning the broader class of cases to which the particular case belonged.

> After this discussion of how we would talk about the operations indicated by exponents, Sam asserted, about the last digit in 5^4, "It *has* to end in a 5." I invited everyone in the class to consider the validity of Sam's decisive assertion and to see if they could explain why he seemed to be so sure. The question I was asking was, How does he know that is true? Harriet said, "Well, anything multiplied by 5 has to end in a 5 or a zero," and Theresa quickly added, "but it has to be a 5 because when you multiply 5 times 5 you get a 5 [for the last digit]." Martha observed, "You times the square number, you square it again and you get 625." And Carl responded, moving to the level of a mathematical generalization, "You don't have to do that. It's easy, the last digit *is always going to be* 5 because [in the standard multiplication algorithm] you are always multiplying last digits of 5, and 5 times 5 ends in a 5." Carl went beyond the question about the last digit in 5^4 and gave both a conjecture and a proof about what must be true of the last digit of *all* of the powers of 5. (At this point in the discourse, the assumed domain for exponents is whole positive numbers.)
>
> Lampert (1990, p. 48) (emphasis in original)

In this episode, Sam made an assertion about the last digit of a specific power of 5, namely 5^4, which essentially gave rise to a proving task involving a single case and serving the purpose of justification: "Prove that the last digit in 5^4 is 5." After Lampert's probing questions and contributions from several students in the class, Sam's idea developed into a generalization that was ultimately expressed by Carl in the form of a conjecture about the last digit in any power of 5: "The last digit of all of the powers of 5 is 5 (all exponents are assumed to be positive integers)." This conjecture essentially gave rise to a new proving task, which involved infinitely many cases and served the purpose of justification. I highlight the fact that none of the two proving tasks in the episode was explicitly posed to the students by the teacher. Rather, both tasks emerged naturally from students' engagement in authentic mathematical activity as they explored what happens to the last digit of powers of 5. Of course, this is not to say that the teacher had not planned, or could not have planned, the students' work on the tasks through deliberate, but "invisible" to the students or to an outside observer, instructional engineering (Stein et al., 2008; Stylianides & Stylianides, 2014b).

To conclude, my focus in this book is on the different kinds of proving tasks in the categorization that I described earlier. Given also my view of proving as being a vehicle to mathematical sense-making (see Chapters 1 and 2), I pay particular attention to proving tasks where the issue of proof is not posed directly or from the outset to students, but rather arises naturally from students' exploration of a mathematical phenomenon or situation.

Characteristics of Proving Tasks That Can Influence Their Generated Proving Activity

The mathematical activity that a mathematics task can potentially generate during its classroom implementation depends on a multitude of factors. Some of these factors relate specifically to the teacher, such as the teacher's knowledge and beliefs (e.g., Collopy, 2003; Corey & Gamoran, 2006; Remillard, 2005; Stylianides & Stylianides, 2008a) or the practices implemented by the teacher when orchestrating class discussion related to a given task (e.g., Stein et al., 2008; Stylianides & Stylianides, 2014b). Other factors relate to the broader instructional context where a task is implemented, such as the amount of time allocated to the task's implementation and the degree of tolerance to errors or risk taking among classroom participants while working on a cognitively demanding task (e.g., Doyle, 1983, 1988; Henningsen & Stein, 1997; Stein et al., 1996).

Notwithstanding the importance of these and other related factors, they are rather generic and can apply potentially to the classroom implementation of a broad range of mathematics tasks. This raises the issue: Do the varying characteristics of the kinds of proving tasks that I described earlier play, or have the potential to play, any specific role in the proving activity that each of these kinds of tasks can generate in the classroom? I propose that the answer to this question is in the affirmative and that different kinds of proving tasks can potentially support qualitatively different proving activities.

Specifically, I propose that the proving activity that a proving task can potentially generate in the classroom depends partly on the following three main characteristics of the proving task itself:

(1) The number of cases involved in the task (a single case, multiple but finitely many cases, or infinitely many cases).
(2) The purpose of the task (to justify or to refute a statement).
(3) The extent to which any conditions of the task are ambiguous and thus subject to different legitimate assumptions by students.

These three characteristics are presented diagrammatically in Figure 3.1.

The first two characteristics derive directly from the categorization of proving tasks that I discussed in the section "A Categorization of Proving Tasks." The third characteristic emerges from the realization that the conditions of a proving task are usually, or can be, subject to different assumptions by students (Herbst & Brach, 2006), with different legitimate assumptions having the potential to give rise to different solution paths and thus variant proving activities (Fawcett, 1938; Stylianides, 2007a). At a basic level, there may be an unintentional ambiguity in the conditions of a proving task due to an oversight in its design or phrasing. There may be cases, though, where an ambiguity is practically unavoidable due, for example, to the tediousness of an alternative task formulation that would endeavor to spell out every single detail so as to filter out all possible, undesirable interpretations of its conditions. At a more substantial level, the design of a proving task with ambiguous conditions can be deliberate and part of a purposeful instructional engineering that aims to

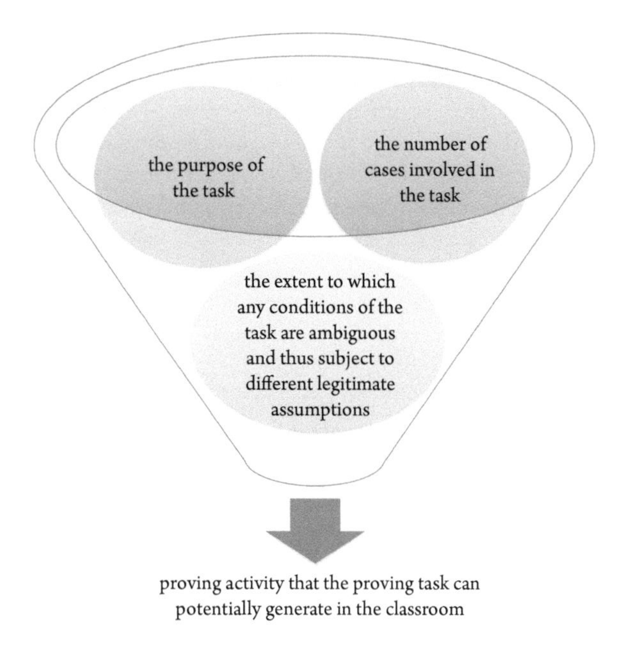

Figure 3.1 The proving activity that a proving task can potentially generate depends partly on three main characteristics of the task itself.

support specific learning goals. For example, a task whose conditions are subject to different legitimate assumptions can offer a useful learning context to enhance students' appreciation of the role that assumptions play as building blocks of arguments, proofs, or mathematical theories (Fawcett, 1938; Jahnke & Wambach, 2013; Stylianides, 2007a).

Proving tasks that vary across the three characteristics are likely to generate qualitatively different proving activities during their classroom implementation, not least because certain modes of argumentation (cf. Table 2.1) are particularly relevant, or generally more relevant than others, to the solution of specific proving tasks. For example, the construction of a counterexample or an argument by *reductio ad absurdum* are more likely to be used, or called to use, during the implementation of tasks whose purpose is to *refute* rather than to *justify* a statement (cf. characteristic 2 in the above list).[3] A further example concerns the systematic consideration of all the cases involved in a situation, which is a mode of argumentation more likely to be used, or called to use, during the implementation of tasks that involve *finitely* rather than *infinitely* many cases (cf. characteristic 1). Also, a proving task with an ambiguous condition that allows different legitimate assumptions about the set of

[3] As I explained in footnote c of Table 2.1, by *reductio ad absurdum* I mean the method of proof that demonstrates a statement is *false* by showing that its acceptance leads to a contradiction. Thus I view *reductio ad absurdum* as a method of *refuting* a statement. A method closely related to *reductio ad absurdum* is *proof by contradiction*, which I defined to be the method of proof that demonstrates a statement is *true* by showing that acceptance of its negation leads to a contradiction. Thus I view proof by contradiction as a method of *justifying* a statement.

cases involved in the task, or about whether the task calls for justification or refutation of a given statement, is more likely than a less ambiguous task to generate a range of modes of argumentation in students' proving activity (cf. characteristic 3). Specifically, different students, at different stages during the proving activity, may produce arguments that span sets of different cardinalities or arguments that aim to achieve different purposes.

There is no doubt that the relationship between proving tasks and corresponding proving activity is complex, and indeed there are exceptions to the general trends I have described. Take, for example, the mode of argumentation associated with the systematic consideration of all the cases involved in a situation. While this mode of argumentation is generally more relevant to tasks concerning finitely many cases, it can also be used in proving tasks concerning statements formulated over infinite sets, such as the following: "Prove that the sum of any two odd numbers is an even number" (the assumed domain of reference is the set of natural numbers). There are different viable definitions of odd and even numbers that one can choose from in trying to prove this statement (see, e.g., Ball & Bass, 2000a, pp. 83–84), and each of these definitions can support the construction of a different kind of argument. Let me consider the so-called units digit definitions, according to which "an odd number is a natural number whose units digit is 1, 3, 5, 7, or 9" and "an even number is a natural number whose units digit is 0, 2, 4, 6, or 8." With these definitions one can prove the statement by considering systematically the sums of all the different combinations of the last digits of two odd numbers, confirming that every such sum yields a natural number with a last digit of 0, 2, 4, 6, or 8, i.e., an even number.[4] In Chapter 2 we saw another proof for this statement, by an elementary school student called Jamie, that used different definitions of even and odd numbers and a different mode of argumentation (Carpenter et al., 2003, p. 90). To further illustrate the complexity of the matter, I consider a proving task that involves a finite number of cases and whose solution could draw, in theory, on a systematic consideration of cases: "How many towers can you create that are 15 blocks tall using only yellow and red blocks? Prove your answer." It would be unreasonable to expect that one would solve this task by enumerating systematically all 32,768 cases; rather, the solver would have to find and apply a viable alternative mode of argumentation for its solution.[5]

To conclude, I am well aware that the relationship between proving tasks and corresponding proving activity defies general rules. Yet I consider it important to explore and try to articulate some aspects of this complex relationship, taking into account also the mediating role of the teacher. The fact that my study of the relationship is restricted to the mathematical work of elementary school students renders the domain less complex, for I presume that the relationship takes (slightly) different forms at different educational levels. Indeed, even if all factors could be kept the same, students' knowledge about proof at the different levels would differ (e.g., their knowledge of specific proof methods such as proof

[4] This argument is based on the premise that the last digit of the sum of two natural numbers is the last digit of the sum of their units digits, which students can justify in the context of the standard written method for addition.
[5] Research with elementary students working on different versions of this particular proving task showed that students might develop alternative modes of argumentation by abstracting basic rules of counting (Maher, Powell, & Uptegrove, 2010).

by contradiction or proof by mathematical induction). The study of the relationship at any level of education can help generate useful knowledge about the use of specific proving tasks in the service of specific learning goals in the mathematics classroom, as well as about different kinds of proving tasks that can offer students opportunities for broad-based and deep learning in the area of proving.

Data Sources and Analytic Method

In preparing this book I had data available from four elementary classes in state schools in two different countries: a third-grade class (8–9-year-olds) in the United States, and two Year 3 classes (7–8-year-olds) and one Year 4 class (8–9-year-olds) in England. The two Year 3 classes were taught by the same teacher, so there were three teachers in total. Across all four classes, the students' ages ranged from 7 to 9 years, roughly covering the middle age band of elementary students internationally.

I first describe the data sources. As part of that description, I also make a case for the appropriateness of the available data given my goals in the book, and I explain my rationale for restricting my examination to data from only two of the available classes. I finish the chapter with a description of the analytic method I followed in my examination.

Data Sources

The teacher of the American class was Deborah Ball, a well-known teacher–researcher (see, e.g., Ball, 1993, 2000) who taught the mathematics lessons in that particular third-grade class for an entire school year. Ball's teaching was carefully documented in a large database of the Mathematics Teaching and Learning to Teach Project (MTLTP) at the University of Michigan. The data included: videotapes and audiotapes of the classroom lessons; observation notes; classroom and interview transcripts; copies of students' work in their notebooks, homework assignments, and quizzes; and copies of the teacher's journal entries with her lesson plans and teaching reflections. In my earlier research (e.g., Stylianides, 2005, 2007a–c; Stylianides & Ball, 2008) I used these data to explore a number of theoretical and empirical issues related to the meaning, teaching, and learning of proving in the elementary mathematics classroom.

In this book I draw on some of the MTLTP data that I used in my previous research, notably, transcripts of classroom episodes and excerpts from the teacher's journal entries. These data, together with the new data that I collected in England (details of which I will give later), formed an expanded dataset in the service of this book's particular focus on the interplay between proving tasks and proving activity and on the mediating role of the teacher. I had originally considered focusing the book entirely on the newly collected data from England, but I withdrew from that idea for two main reasons: using data from Ball's class alongside classroom data from England (1) offered me the opportunity to explore teaching in two different countries with the potential to reveal issues that could have otherwise remained salient and (2) allowed me to consider insights from my investigation of the familiar MTLTP data against fresh insights from my investigation of the newly collected data. This complementarity of research contexts and insights has also created a favorable setting for synthesis, extension, and further development of my previous research on proving in the elementary mathematics classroom.

Ball's class was socioeconomically, ethnically, and racially diverse, with 22 students of different achievement levels. The mathematics lesson period in the class was approximately an hour, 5 days a week. During each period, the class worked on one or two tasks, carefully selected or designed by the teacher to be mathematically rich. The period often began with the students exploring a task individually or in pairs, then in small groups, and ultimately in the whole group. The curriculum was organized around units on general mathematical topics, such as number theory, integer arithmetic, and probability.

Ball's teaching of the third-grade class was organized as a year-long teaching experiment. One of the goals of this experiment was to explore what it would mean and look like to help young children become skilled mathematical reasoners. To promote this goal, Ball modeled her "classroom as a community of mathematical discourse, in which the validity for ideas rest[ed] on reason and mathematical argument, rather than on the authority of the teacher or the answer key" (Ball, 1993, p. 388). Inspired by Bruner's (1960) notion of *intellectual honesty*, Ball's teaching was a continuous struggle to achieve a defensible balance between the two (often competing) imperatives that I discussed in Chapter 2: *mathematics as a discipline* and *students as mathematical learners*. Ball noted the following about her pedagogical practice in one of her writings:

> I must consider the mathematics in relation to the children and the children in relation to the mathematics. My ears and eyes must search the world around us, the discipline of mathematics, and the world of the child with both mathematical and child filters.
>
> Ball (1993, p. 394)

Partly because of Ball's goal of helping her students to become skilled mathematical reasoners, issues of proving were prominent in her class. Issues of task design and implementation were also prominent in the available classroom data because, as I mentioned earlier, Ball organized each lesson around one or two mathematics tasks, which she either selected or designed to serve the lesson objectives. Thus Ball's pedagogical practice offered a rich and well-documented context where all of the main issues of interest to this book emerged and could be researched. It is important to note, however, that Ball did not have in mind, as she was teaching or planning for her teaching, the definitions of proof and proving that I described in Chapter 2; I developed these definitions years after the completion of the teaching experiment (Stylianides, 2007b). The same applies for the relationship between proving tasks and proving activity that I explore in this book.

There is no doubt that Ball's pedagogical practice is not typical, but that is not a problem for my purposes in this book as I am not focusing on the place of proving in ordinary or representative elementary classrooms. Indeed, such a focus would have been of little interest to the field, for, as I discussed in Chapter 2, it is common knowledge among researchers and practitioners internationally that proving currently has a marginal place in most elementary mathematics classrooms. Following recognition that this state of affairs is problematic (cf. Chapter 2), a key question that arises, and which this book takes a step to address, is this: What would it take for elementary teachers to productively engage their students in proving? From a methodological standpoint this question calls for examination of non-typical pedagogical practices, such as those of Ball, Lampert (1992, 2001), and Zack (1997), which took issues of proving seriously and used appropriate mathematics tasks to support their students' learning of proving. Examination of these sort of deviant, information-rich pedagogical practices affords researchers an opportunity to better understand what is involved in trying to elevate, by means of proving tasks, the place of proving in elementary students' mathematical work. I view this as the first, foundational, step in an ambitious, long-term research program that would aim to help other elementary teachers to also offer their students productive learning opportunities with proving.

The teachers of the English classes, whom I call Mrs. Howard and Mrs. Lester (pseudonyms), were experienced, sought to help their students become skilled mathematical reasoners, and were recognized in their local contexts as being competent teachers of elementary school mathematics. An indication of their recognition as competent teachers is that, during the year I was collecting data for the book, both of them spent approximately 1 day a week on secondment to a local university where they contributed to the mathematics strand of the university's elementary teacher education program. The program was based on research that highlighted the importance of bridging university-based and school-based perspectives in the preparation of teachers (see, e.g., Hagger & McIntyre, 2000), and it was thus organized as a partnership between the university's department of education and local schools where prospective teachers had their field placements. The pedagogical practices of Howard and Lester were not typical, but they were nevertheless appropriate for study in this book for the same reasons that I gave previously for Ball's practice.

My data collection in the English classes happened over a number of lessons spread throughout an entire school year. During that time Howard was the regular teacher of a Year 4 class of 30 students in a large elementary school, which served a socially diverse community close to the center of a small city. Lester was semi-retired at the time and did not have full responsibility for a class but did teach occasionally in different classes at her school, which was relatively small and served the socially diverse community of a few nearby villages. For the purposes of my data collection, Lester taught some mathematics lessons in two Year 3 classes at her school, which had 25 students each and whose regular teachers were newly qualified for teaching (both of them were in their first year of service). Lester's lessons offered an opportunity for in-school professional development of the two beginning teachers who, like me, had an opportunity to observe Lester teaching and discuss with her the planning and implementation of the observed lessons. In both schools most students were from white British backgrounds, with the remaining students coming from a range of minority ethnic groups. Also, the proportion of students with special educational needs was roughly at the national average. Overall, both Howard and Lester taught in ordinary state schools, which were both rated as "good" in their most recent inspections conducted by the Office for Standards in Education (Ofsted).[6]

[6] There were four possible Ofsted inspection outcomes, of which "good" was the second best: "outstanding," "good," "requires improvement," and "inadequate." According to the Ofsted framework, a "good" school was considered to be effective in delivering outcomes that provided well for all its students' needs, ensuring that students were well prepared for the next stage of their education.

The data from Howard's and Lester's classes comprised the following: detailed fieldnotes that I took during my observations of lessons with the teachers engaging their students in proving tasks; copies of the written work produced by the students during these lessons; and notes from my informal interactions with students during the lessons and with the teachers both before and after the lessons. The teachers designed or selected independently virtually all of the proving tasks that they implemented during the lessons I observed. My input to the teachers' choices was broadly limited to a general and rather vague request for them to implement tasks that required students to engage in "proving," which I explained to them as the activity in which students would reason mathematically, make conjectures, explain or justify assertions, etc. Also, across the different lessons that I observed in the three classes, I asked the teachers whether they could use proving tasks that involved sets of different cardinalities. To that end, I offered to the teachers a few examples of tasks in each category (a single case, multiple but finitely many cases, and infinitely many cases) so that they understood what I meant. It is important to note further that, similar to Ball, the two English teachers did not know, and thus did not use in their planning or teaching, the definitions of proof and proving that I described in Chapter 2. I did not consider it appropriate to impose upon them my own definitions of proof and proving, though of course I was interested to see whether these definitions offered a useful lens for my analysis of the teachers' practices, like they did in my previous research of Ball's teaching practice (e.g., Stylianides, 2007a–c).

The two teachers ended up designing or selecting some proving tasks for the first time during their teaching careers, but they also used some (modified versions of) tasks that they had used again in previous years. The design or selection of new proving tasks was motivated partly by the needs of my data collection and partly by the teachers' own interest in expanding their banks of proving tasks. The latter was related to the recent release of the new English national curriculum in mathematics (Department for Education, 2013), which, as I explained in Chapter 2, set for students of all ages a core aim specifically related to proving. As this was one out of only three core aims set by the national curriculum for students of all ages, the two teachers considered it to be a new priority area for the mathematics provision in their schools.

The data from the English classes helped expand and diversify the set of classroom episodes and pieces of student work I could draw on in preparing the book. Also, as part of the data collection I was physically present during lessons when the two teachers implemented proving tasks in their classrooms, and this gave me first-hand experience of observing what might be involved in engaging elementary students with proving.

My plan up until well into the process of writing the book had been to use data from all three English classes, alongside data from Ball's class. Yet this turned out to be unrealistic and in competition with my efforts to keep the book within a reasonable length and the sources of data used across chapters consistent. I thus decided, to my regret, to use data from only one English class: Howard's Year 4 class. The main reason for this choice was that, unlike Lester and similar to Ball, Howard was the regular teacher of that particular class for the entire school year, and so the lessons I observed were fully integrated into students' regular mathematical work.

Analytic Method

My analysis focused on all the episodes that I had available from the two classes, Ball's third-grade class and Howard's Year 4 class, where students engaged with proving tasks. In order to prepare the selection of episodes for the book, I followed the following two-step procedure to organize all proving tasks that were used in the episodes. First, I identified all proving tasks whose conditions were ambiguous and subject to different legitimate assumptions by students (see characteristic 3 of proving tasks described in the section "Characteristics of Proving Tasks That Can Influence Their Generated Proving Activity"). I considered it important to keep these tasks separate from the rest, because, depending on one's interpretation of the conditions of an ambiguous task, the task can be classified differently with respect to the number of cases involved in it (characteristic 1) or its main purpose (characteristic 2). Second, I organized the remaining proving tasks according to whether they involved a single case, multiple but finitely many cases, or infinitely many cases (characteristic 1). I used characteristic 1 rather than characteristic 2 to organize the remaining tasks because some proving tasks evolved through a process of conjecture and revision during their classroom implementation, and so often the purposes of justification and refutation both came into play within the context of the same task.

This procedure helped organize the proving tasks in the episodes into the following four groups: proving tasks with ambiguous conditions; proving tasks involving a single case; proving tasks involving multiple but finitely many cases; and proving tasks involving infinitely many cases. Both teachers tended to implement proving tasks of the same kind within a lesson or over a sequence of lessons, which offers suggestive evidence that the particular organization is not only meaningful mathematically but also sensible from the point of view of teaching practice.

For each of the four kinds of proving tasks, I selected a pair of episodes, one from each class, where the students engaged with one or more proving tasks belonging to the particular kind (i.e., I selected a total of eight episodes). I selected the specific episodes in each pair for two reasons. First, for their illustrative power: They described rich classroom activity in the context of the particular kind of proving task, thus offering vivid images of what might be involved for the teacher and the students in an elementary classroom during the implementation of these tasks. Second, for their complementarity: They raised some common issues that helped reveal patterns of student proving activity or teacher actions in the context of the particular kind of proving task, but also some different (and at times contrasting) issues that helped illuminate points that could have passed unnoticed if the analysis had been limited to one episode.

To prepare the selected episodes for analysis, I created detailed descriptions of the episodes using, and triangulating among, all of the data sources that were available to me, but primarily the lesson transcripts (based on the available video records) of the episodes from Ball's class and my fieldnotes of the episodes from Howard's class. The episode descriptions were a combination of narration of classroom events, dialogues between classroom participants, and presentation of pieces of written work produced by students during the lessons. Due to the absence of video records of Howard's lessons (dictated by the permission

I received to conduct research in her class), it is possible that there are some differences between the presented dialogues and what was actually said by classroom participants during the lessons. In cases where I was unsure about the accuracy of my notes of participants' actual words, or when I did not manage to capture fully those words, I presented them in paraphrase mode, communicating the meaning of what was said. I showed to Howard the descriptions of all the episodes from her class that I selected for use in the book, and she confirmed to me that the descriptions matched her memories of the lessons. All student names in the descriptions of episodes from both classes were replaced with pseudonyms.

In each episode, as well as across the two episodes in the same pair, I first analyzed mathematically the proving tasks used by the teachers so as to establish their structural similarities, but also to identify any notable differences between them (e.g., in their choice of numbers, context, or overall formulation). The identification of any notable differences between proving tasks of the same general kind was useful in trying to explain, later in the analysis, any notable differences in students' proving activity as they engaged with these tasks.

In each episode, and across the episodes in the same pair, I further analyzed students' proving activity as they engaged with proving tasks of the same kind. In the particular case of tasks with ambiguous conditions, I analyzed students' interpretations of the conditions of the tasks and the arguments that students developed based on different interpretations or assumptions. My analysis of students' arguments for this and all other kinds of proving tasks was based on the definition of proof that I discussed in Chapter 2, including the following three components into which the definition breaks any given mathematical argument: the set of accepted statements, the modes of argumentation, and the modes of argument representation. The analysis allowed me to see whether students' arguments met the standard of proof, and if not what component or components of an argument seemed to be lacking or causing difficulty to students. The mathematical analysis of proving tasks as I have described, together with the findings of the relevant literature, offered me a useful lens through which I tried to explain any general difficulties students had with certain kinds of proving tasks or any variations in students' difficulties during their engagement with proving tasks of the same kind.

Finally, I analyzed the role of the teachers as they managed their students' work on proving tasks. In particular, I analyzed what a teacher's rationale was, or might be, for the choice or sequencing of the specific proving tasks in an episode; the teacher's actions or (presumed) decisions when students shared their arguments, especially when these arguments did not meet the standard of proof, as well as the implications of those actions or decisions for students' subsequent proving activity. A cross-episode comparison of what was involved for the two teachers during the implementation of the same kind of proving task allowed some general aspects of a teacher's role to emerge, including challenges for teachers as they manage students' proving activity in the context of those tasks.

Overall, the analysis was not intended to yield any generalizable findings, but rather to help deepen understanding of the relationship between proving tasks and proving activity, and of the mediating role of the teacher. The organization of the next four chapters maps onto the organization of the analysis around the four kinds of proving tasks. Specifically, each chapter focuses on one kind of proving task, with the discussion being situated in the context of the selected pair of classroom episodes: Chapter 4 focuses on proving tasks with ambiguous conditions, while Chapters 5–7 focus on proving tasks that involve, respectively, a single case, multiple but finitely many cases, and infinitely many cases.

4

Proving Tasks with Ambiguous Conditions

In this chapter I examine proving tasks whose conditions are ambiguous and thus subject to different legitimate assumptions by students. The emphasis is on the proving activity that this kind of task can help generate in the classroom and on the role of the teacher while implementing the tasks. My discussion is situated in the context of two classroom episodes where tasks of this kind were used: the first (Episode A) comes from Howard's Year 4 class in England (8–9-year-olds); the second (Episode B) comes from Ball's third-grade class in the United States (again 8–9-year-olds). These episodes have been selected for their illustrative power and for the complementary issues they raise about what it might mean or look like when elementary teachers use in their classes proving tasks with ambiguous conditions. I describe and discuss each episode separately, and I conclude with a more general discussion.

Episode A

Description of the Episode

This episode took place in Howard's Year 4 class in a lesson that Howard taught on October 1. Her main goal for this lesson was to engage her students in mathematical reasoning and problem solving in the context of number sentences that equal 10, also practicing addition and subtraction facts. During the lesson she used one main task, which I call Task 1, with three variants of this task, Tasks 1a–c, and one extension of the last variant, Task 1c′. I have numbered the tasks in this way to indicate their relationships and for ease of reference when I discuss the episode in the next section ("Discussion of the Episode").

The lesson started with Howard presenting the following task to the class:

1. *Task 1*: "How many ways do you think there are to make the number 10? (Prove your answer.)"

Proving in the Elementary Mathematics Classroom.

Although the last part of the text in parentheses was not an explicit part of the statement of the task, it was understood in the class, and became clear during the task's implementation, that the teacher not only expected students to answer the question (i.e., to specify the number of ways they thought there were), but also to prove their answer.

The students started working on the task individually or in pairs, and they wrote different kinds of number sentences, as illustrated by the written work of three students presented in Figure 4.1.[1] Across these number sentences, we see examples of all four operations being used, some commutative pairs of number sentences, and a few number sentences involving decimal numbers.

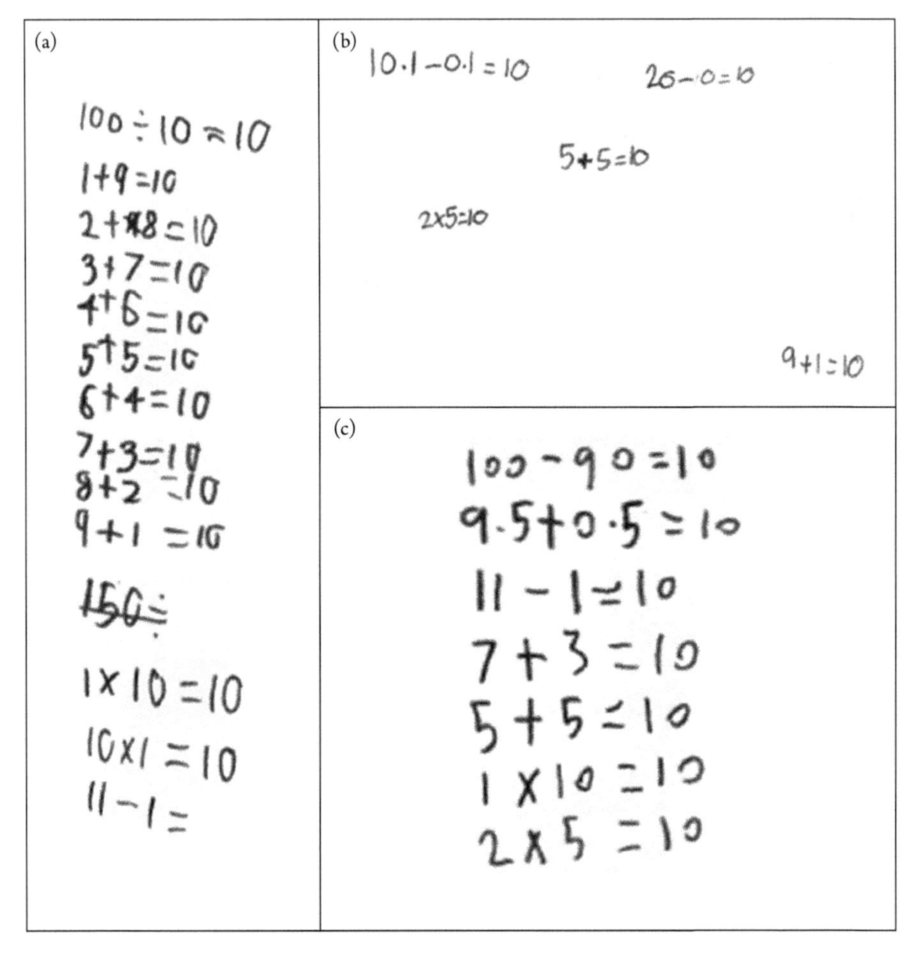

Figure 4.1 Samples of student work in response to Task 1.

[1] It was not possible to identify which students produced which work, as the teacher had not asked students to write their names on their papers.

After a few minutes, Howard brought the class together to discuss questions that some students had asked her regarding the conditions of the task. Clara asked first whether commutative pairs of number sentences could count as being different:

2. *Clara*: Can you take them [the numbers] round?
3. *Howard*: What do you mean?
4. *Clara*: Can you switch them round, like $9 + 1 = 10$ and $1 + 9 = 10$?

Howard invited comments from the rest of the class. Osborn said:

5. *Osborn*: These are the same digits. It's not a different calculation.

Following a brief discussion, the class concluded that commutative pairs of number sentences were essentially presenting "the same thing" and thus it did not make sense to say that they offered different ways to make the number 10.

Simon then asked a new question about the conditions of the task:

6. *Simon*: Is it just adding, or you can do also subtraction, multiplication, and division?

The class agreed that different operations should be allowed as each of them helps generate different number sentences. Andy then observed the following:

7. *Andy*: That would be never ending! If you can take away, you can take on forever!
8. *Howard*: How do we call "never ending"?
9. *Another student*: Infinity.
10. *Howard*: Can you write down on your pieces of paper to show the subtraction calculations that can take you on forever?

The students worked individually or in pairs on Howard's prompt. Figure 4.2 presents the work of two students.

In the whole class discussion that followed, Andy presented his work on the board.

11. *Andy's writing on the board*: $10 - 0$
$11 - 1$
$12 - 2$
$13 - 3$
$14 - 4$
$15 - 5$
$16 - 6$
$17 - 7$
$18 - 8$

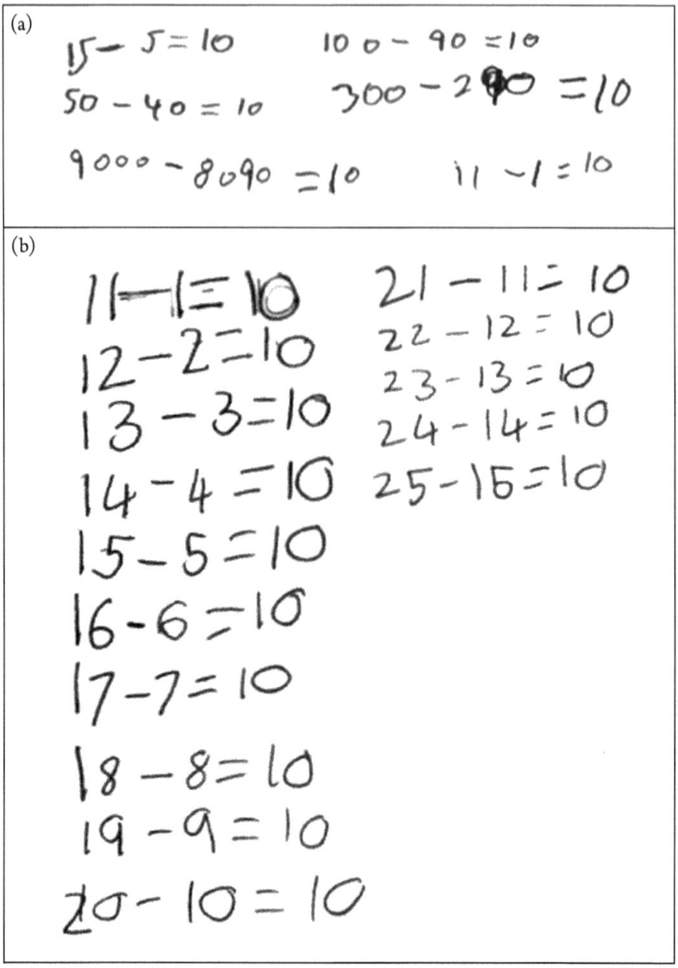

(a)

15 − 5 = 10   100 − 90 = 10
50 − 40 = 10   300 − 290 = 10
9000 − 8090 = 10   11 − 1 = 10

(b)

11 − 1 = 10   21 − 11 = 10
12 − 2 = 10   22 − 12 = 10
13 − 3 = 10   23 − 13 = 10
14 − 4 = 10   24 − 14 = 10
15 − 5 = 10   25 − 15 = 10
16 − 6 = 10
17 − 7 = 10
18 − 8 = 10
19 − 9 = 10
20 − 10 = 10

Figure 4.2 Samples of student work intended to illustrate that there are an infinite number of subtraction number sentences for 10.

Howard invited the class to say what they noticed about Andy's work. A student observed that in each case Andy was subtracting the units digit of the first number. Another student observed that Andy could go on forever writing new number sentences simply by adding 1 to each term in the previous sentence. This procedure is more clearly illustrated in Figure 4.2(b), which includes examples where the subtrahend has more than one digit.

Howard confirmed that by using subtraction one could write as many number sentences as one wanted to. She then referred to the question that Simon had raised earlier (line 6), and she asked the students to limit their work to *addition* number sentences. With what she just said, and given also what the class had agreed earlier about commutative pairs of number sentences not counting as different, the teacher essentially gave to the class the following variant of the original task.

12. *Task 1a*: "How many ways do you think there are to make the number 10 if we just use addition and commutative pairs of number sentences do not count as different? (Prove your answer.)"

The students' work on this task included again a variety of number sentences, including sentences with negative or decimal numbers as illustrated in the samples of student work presented in Figure 4.3.

After some time when students worked on Task 1a individually or in pairs, Howard brought the students together for a whole class discussion of questions she had received from several students about the conditions of the task. Irene raised a question about the number of addends in each number sentence:

13. *Irene*: Can you add three numbers, say 1 + 5 + 4?

(a)	(b)
0+10=10	
10+0=10	−1+11=10
−1+11=10	−2+12=10
−2+12=10	−3+13=10
−3+13 is 10	−4+14=10
−4+14 is 10	−5+15=10
−5+15=10	−6+16=10
−6+16=10	−7+17=10
9.9+0.1=10.0	−8+18=10
9.8+0.2=10.0	
9.8+0.3=10.0	

Note: the two number sentences that are a bit difficult to read are as follows:

−3 + 13 *is* 10
−4 + 14 *is* 10

Figure 4.3 Samples of student work in response to Task 1a.

Simon followed up on this with a further question about whether the addends could be negative numbers:

14. *Simon*: Could you say $-10+20=10$ or $-20+30=10$?
15. *Howard*: How many ways would there be if you were allowed to use negative numbers?
16. *Simon*: Infinitely many!

The teacher then clarified which conditions she wanted the class to assume when solving the task:

17. *Howard*: I don't want you to use multiple addends, just two numbers, and I don't want you to use negative numbers. How many ways are there?

Howard essentially gave to the class a new variant of the original task, which I call Task 1b:

18. *Task 1b*: "How many ways do you think there are to make the number 10 if we just use two non-negative addends and commutative pairs of number sentences do not count as different? (Prove your answer.)"

The students then worked on Task 1b individually or in pairs. Most students used number sentences that combined whole numbers (i.e., non-negative integers) and positive decimal numbers involving tenths, as illustrated in the samples of student work in Figure 4.4.

Howard circulated around the class, observing students' work and engaging them in dialogue about their work. After a brief chat with Keith, Howard called for a new whole class discussion relating to Keith's work. Howard asked Keith to explain to the rest of the class what he had done.

19. *Keith*: I looked at using decimal numbers.
20. *Howard*: What did you think [about the number of ways there would be then]?
21. *Keith*: There may be quite a few when you complete the tenths. [He wrote this number sentence on the board: $9.9+0.1=10$.]

Howard invited comments on Keith's work from other students. Orrin and Osborn said that they had also been using decimal numbers. Osborn said further:

22. *Osborn*: I think there may be 105 ways, because there are a hundred different decimal points.

Howard wrote on the board the number 9.9 and above its tenths digit she wrote the letter "t." Then she asked the class what would come after the tenths digit.

23. *Orrin*: Is it negative?

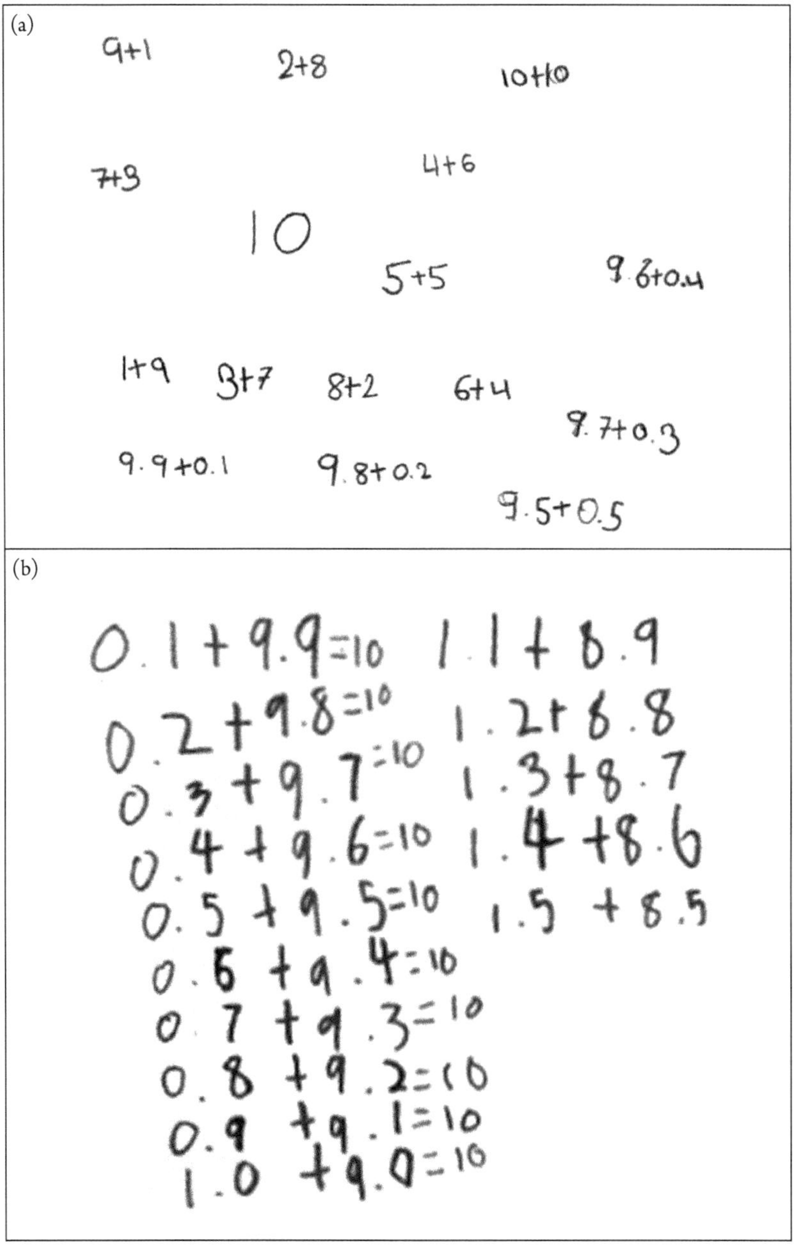

(a)

9+1 2+8 10+10

7+3 4+6

10

5+5 9.6+0.4

1+9 3+7 8+2 6+4

9.7+0.3

9.9+0.1 9.8+0.2

9.5+0.5

(b)

0.1 + 9.9=10 1.1 + 8.9
0.2 + 9.8=10 1.2 + 8.8
0.3 + 9.7=10 1.3 + 8.7
0.4 + 9.6=10 1.4 + 8.6
0.5 + 9.5=10 1.5 + 8.5
0.6 + 9.4=10
0.7 + 9.3=10
0.8 + 9.2=10
0.9 + 9.1=10
1.0 + 9.0=10

Figure 4.4 Samples of student work in response to Task 1b.

Several students agreed with Orrin. In a discussion that I had with Howard after the lesson, she said she had not expected students to have these difficulties with decimal numbers. However, she could not have tried to address these difficulties during the lesson because that would have been a major deviation from her plan. Thus she proceeded by asking the students to exclude decimal numbers from their calculations.

24. *Howard*: We will think more about decimals in another lesson. For now we will use no decimals in solving the task.

With this decision, Howard essentially gave to the class yet another variant of the original task, which I call Task 1c:

25. *Task 1c*: "How many ways do you think there are to make the number 10 if we just use two whole number addends and commutative pairs of number sentences do not count as different? (Prove your answer.)"

After the students worked on Task 1c individually or in pairs for a few minutes, Howard brought the class together for another discussion. The discussion started with Jim presenting his work. Jim wrote his list of number sentences on the board as follows:

26. *Jim's list*: $1+9$
$2+8$
$3+7$
$4+6$
$5+5$

Then Jim commented:

27. *Jim*: Switching numbers round doesn't give different ways.

Another student proposed that the number sentence $0+10$ should also be in the list; Howard wrote this number sentence on the top of Jim's list.

28. *Jim's expanded list*: $0+10$
$1+9$
$2+8$
$3+7$
$4+6$
$5+5$

Howard then asked the students in the class to raise their hands if they had written their number sentences "in order," like Jim did. About half of them said they had done that. Howard invited Harriet, one of the students who said they had not written their number sentences in order, to tell the class the specific order in which she had written the number sentences. The teacher wrote Harriet's list on the board.

29. *Harriet's list:* $2+8$

$1+9$

$3+7$

$4+6$

$5+5$

Howard then asked the students to compare the two lists and say what they noticed. Howard first called on Pam who, after taking some time to think, said she "didn't know." Several other students contributed to the discussion, observing that the "ordered list" (as some of them called Jim's list) was easier to read than the "mixed up list" (as some of them called Harriet's list). They observed further that the ordered list helped one keep track of which number sentences had already been written. When Howard asked the students to say why this feature of an ordered list was "a good thing," two students responded as follows:

30. *Rania*: Because you may do one twice.
31. *Another student*: You might have missed one.

Howard repeated these two reasons and highlighted the importance of students following a systematic way when they wanted to list all possible ways in a problem. Then she asked the students to work on an extension of Task 1c:

32. *Howard*: Choose another number between 10 and 20, not 10, and use two whole numbers and just adding—no negative numbers, no decimals, no switching around. How many ways are there?

Several students asked questions to confirm what Howard had just said about the conditions of the new task: that the number sentences should involve only two whole number addends and that commutative pairs of number sentences did not count as different. This task could be phrased as follows (I call this Task 1c′ so as to indicate that it is basically an extension of Task 1c):

33. *Task 1c′*: "Choose a whole number from 11 to 20. How many ways do you think there are to make this number if we just use two whole number addends and commutative pairs of number sentences do not count as different? (Prove your answer.)"

The students worked individually or in pairs on their chosen numbers. At some point Howard asked them to choose other whole numbers from 11 to 20 and find again all the ways for each of them.

A few minutes later Howard brought the students together for a whole class discussion. The review of students' work began with the number 10, which had been discussed earlier during the lesson (in the context of Task 1c).

34. *Howard*: How many ways did you find for 10?
35. *Pam*: Sixteen.

Howard asked Pam to say all the ways, one at a time, and ticked them off from the list that was already on the board from earlier during the lesson (line 28). When Pam had difficulty offering more than six ways, Howard asked her to say why she thought that was the case. Pam said:

36. *Pam*: I haven't done them in order.

Howard then invited Andy to say how many ways he had found for 10, and he said six. The rest of the class agreed with Andy and Howard recorded the answer.

Howard then asked whether anyone had worked on number 20, and a student said she had found 11 ways for it. Several other students agreed with that and the answer was recorded. The next number was 19. Emma said she had found ten ways for it; there was no disagreement with that and again the answer was recorded. The next number was 14 and Zenia said she had found five ways for it. The following dialogue then happened between Howard and Zenia:

37. *Howard*: Do you think there are fewer ways [to make 14] than there are for 10?
38. *Zenia*: [pause] No.
39. *Howard*: I'll leave you to think about it.

The next number that the class discussed was 18. Jack said he had found ten ways for it; the answer was recorded, as there were no objections to it from the rest of the class. Howard then pointed at the answers recorded thus far, and she asked the students:

40. *Howard*: What do you notice?
41. *A student*: The number of ways goes down as the number gets smaller.

Howard repeated what the student had just said and asked the class to work on a few more numbers and look for patterns. As the students worked individually or in pairs, Howard was circulating around in class, asking students questions about their approaches to finding all ways (systematic versus non-systematic) and about any pattern they saw emerging from their results. Figure 4.5 presents the written work of three students at this point during the lesson.

A few minutes before the end of the lesson, Howard brought the students together for a final whole class discussion. The following results were added to their previous list: six ways for 11; seven ways for 12; and seven ways for 13. Then Howard asked the class:

42. *Howard*: How many ways do you think for 14?
43. *Harriet*: Eight.
44. *Howard*: Why?
45. *Harriet*: I noticed a pattern.

A similar discussion happened about the number of ways for 15, with Andy saying the following:

46. *Andy*: [I think there will be] eight ways because it goes: six, six, seven, seven, eight, eight. [He referred to the ways for 10, 11, 12, 13, 14, and 15, respectively.]

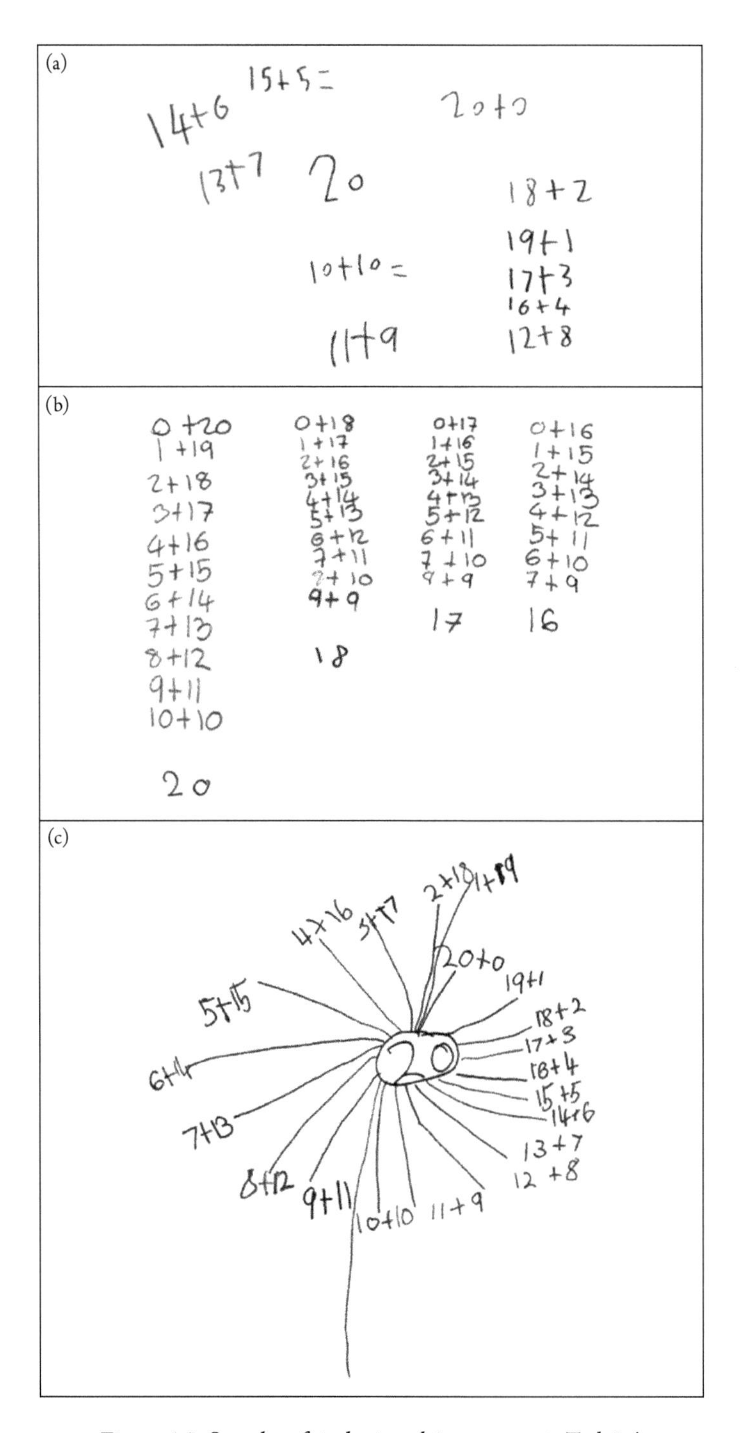

Figure 4.5 Samples of student work in response to Task 1c′.

The class period finished with Howard saying the following:

47. *Howard*: I wonder whether we could say how many ways there are for *any* number? That's for another time.

Discussion of the Episode

The main proving task with which Howard engaged her students in this episode, Task 1 (line 1), was purposefully ambiguous. The points of ambiguity related to the following task conditions about which no information was offered in the statement of the task: the operation(s) one could use in generating number sentences for 10 (addition, multiplication, etc.); the number domain(s) in which the terms of a number sentence could belong (integers, decimals, etc.); the number of terms that a number sentence could have; and which number sentences counted as "different." The teacher engaged the students in proving activity in the context of this and three variants of the main task, Tasks 1a–c (lines 12, 18, and 25, respectively), with the task conditions being discussed and clarified gradually during the lesson, culminating in an unambiguous task statement in the last variant (Task 1c) and its extension (Task 1c′, line 33).

In all of these tasks, students engaged in proving activity leading primarily to *justification* of claims, though there was some variation across tasks in terms of both the number of cases that the proving activity had to consider and the actual claim that the activity had to justify. On the one hand, the conditions of the main task and its first two variants allowed for consideration of an *infinite* number of possible cases, with the corresponding proving activity leading to the conclusion that there were infinitely many ways to make the number 10. On the other hand, the conditions of the last variant of the main task restricted the possibilities to a *finite* number of cases, with the corresponding proving activity leading to the conclusion that there were six ways to make the number 10. Similar to the last variant, its extension also restricted the possibilities to a finite number of cases, with the corresponding proving activity leading to the conclusion that there were a finite number of ways to make each whole number from 11 to 20.

Figure 4.6 summarizes the work of the class on the main proving task and its three variants. In what follows I discuss further the work of the class in Episode A, paying particular attention to the role that the various task conditions played in that work.

Tasks 1, 1a, and 1b: An infinite number of cases

As the lesson unfolded, students themselves identified points of ambiguity in the conditions of the main proving task and its first two variants: In relation to Task 1, Clara asked whether commutative pairs of number sentences could be considered as different (lines 2 and 4) while Simon asked which operations were allowed to be used in the number sentences (line 6). In relation to Task 1a, Irene asked about the number of terms that a number sentence could have (line 13), while Simon asked whether these terms could be negative numbers (line 14). In relation to Task 1b, Keith asked whether the terms of a number sentence could be decimal numbers (line 19). The teacher engaged the class in discussion of

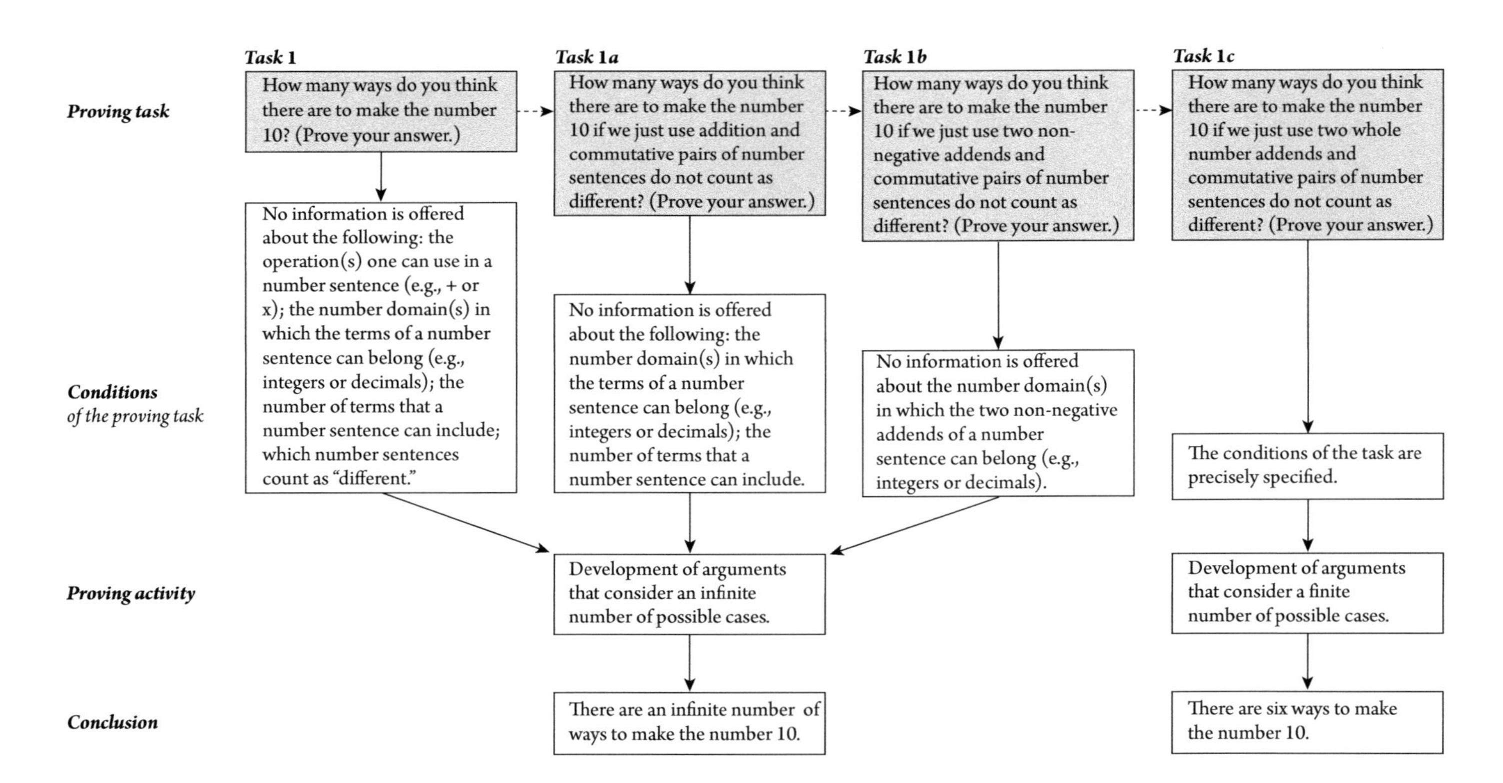

Figure 4.6 Summary of the work of the class on the main proving task and its variants in Episode A.

these questions and of their implications for the answer to the corresponding task, concluding each time with a statement of the conditions that she wanted the students to assume in their subsequent work.

Even though the task conditions from Task 1 to Task 1a, and from Task 1a to Task 1b, introduced further restrictions to the permissible number sentences for 10, they all allowed for an infinite number of ways to make the number 10. The observation that there were indeed an infinite number of ways to make 10 was made and discussed in the context of all of these tasks, with some interesting arguments offered or alluded to by students.

In Task 1, Andy observed that allowing subtraction number sentences would result in an infinite number of ways to make 10 (lines 7–9); Howard followed up on that observation (line 10), essentially asking the students to come up with a procedure for generating subtraction number sentences for 10 that could run *ad infinitum*. Andy's work (line 11), together with the class discussion that followed up on that work, helped describe the following procedure, which is more clearly illustrated in Figure 4.2(b): start with the subtraction number sentence $10-0=10$; for each new number sentence, add 1 to both the minuend and the subtrahend of the previous number sentence.[2]

The work of the class on describing this procedure in the context of Task 1, with contributions from Andy and other students, could be considered as offering a proof for the claim that there are infinitely many number sentences for 10. This is because the argument underpinning that work had the following characteristics, which satisfied the definition of proof discussed in Chapter 2:

(1) It used true statements assumed to be known to the community, notably that the difference between two numbers is preserved if the same number is added to both the minuend and the subtrahend.

(2) It employed (implicitly or explicitly) valid modes of argumentation that were, presumably, within the conceptual reach of the community, notably, the inference rule, "To show that there are an infinite number of number sentences for 10, it suffices to show that a subset of these sentences has infinite cardinality."

(3) It was communicated appropriately using a combination of verbal and written language that the community could easily understand.

Task 1a excluded subtraction number sentences, but it did allow the use of negative numbers as addends in the number sentences. The observation for the existence of an infinite number of ways to make 10 was made by Simon after he had been probed by Howard to say how many ways there would be if one used negative numbers (lines 14–16). The class discussion around Task 1a does not allow us to pinpoint a specific instance when a student described a procedure for generating number sentences for 10 that could run *ad infinitum*. However, the examples of number sentences that Simon offered (line 14) appeared to have satisfied the teacher that Simon had figured out such a procedure: start with the addition number sentence $-10+20=10$; for each new number sentence, subtract 10 from the first

[2] This procedure gives rise to the following infinite set of number sentences for 10: $\{(10+n)-(0+n)=10, \text{ for all } n\in\mathbb{N}\}$.

addend and add 10 to the second addend of the previous number sentence.[3] The student's work that is presented in Figure 4.3(b) implies a slightly different procedure, which also met the conditions of Task 1a: start with the addition number sentence $-1+11=10$; for each new number sentence, subtract 1 from the first addend and add 1 to the second addend of the previous number sentence.[4]

Each of these two procedures in the context of Task 1a could very well form the basis of a proof for the claim that there were infinitely many number sentences for 10. The reasoning for that proof would be similar to the proof that I discussed earlier in the context of Task 1. The only notable difference between the two proofs would be in the set of accepted statements: The proof for Task 1a would presume that the community knew, or could readily accept, that a two-addend sum would be preserved if the same number was subtracted from one of the addends and was added to the other.

The work of the class on Tasks 1 and 1a drew on, and presumably helped solidify or deepen, students' knowledge of subtraction of whole numbers and addition of two integers with opposite signs. The conditions of Task 1b pushed students to further consider, alongside addition of two non-negative integers, addition of two non-negative decimal numbers, as illustrated in the samples of student work in Figure 4.4. Howard had expected that students would be able to use decimal numbers to describe a new procedure for generating number sentences for 10 that could run *ad infinitum*, thus producing a new proof for the claim that there were infinitely many number sentences for 10. Yet students' fragile knowledge of decimal numbers created a major obstacle to this work, which was practically impossible to overcome during the lesson. The students seemed to be able to generate and work with positive decimal numbers involving only tenths (see, e.g., lines 21 and 22, as well as Figure 4.4a, b), but several of them thought that there was a negative number after the tenths digit (line 23). Osborn made an interesting observation (line 22), which implied that he thought there would be a hundred different two-addend number sentences for 10 using decimal numbers involving only tenths. Figure 4.4(b) shows the first few number sentences that would be included in this list: ten number sentences from $0.1+9.9=10$ up to and including $1.0+9.0=10$, with the first addend in each new sentence being one-tenth more than the corresponding addend in the previous sentence. Systematic application of this procedure would indeed generate a hundred different two-addend number sentences for 10, though it is unclear whether Osborn had realized that the last sentence in the list would have to include zero as the second addend (i.e., $10.0+0=10$).

Tasks 1c and 1c′: A finite number of cases

Students' difficulties with decimal numbers prompted Howard to proceed with the introduction of a new restriction to the permissible number sentences for 10, namely that students could only use two whole number addends. This led to Task 1c, the conditions of which were precisely specified and involved only a finite number of possible cases. The

[3] This procedure gives rise to the following infinite set of number sentences for 10: $\{(-10-10n)+(20+10n)=10$, for all $n \in \mathbb{N}\}$.

[4] This procedure gives rise to the following infinite set of number sentences for 10: $\{(-1-n)+(11+n)=10$, for all $n \in \mathbb{N}\}$.

proving activity shifted, then, from a search of procedures for generating number sentences for 10 that could run *ad infinitum* (as in Tasks 1, 1a, and 1b) to a search for ways that would help demonstrate that all different, and finitely many, number sentences were found. Students' work thus far had shown that some, but not all, of them were able to follow a systematic way to write down number sentences of a certain kind (compare, e.g., the samples of students' work presented in Figures 4.2b and 4.4b with those presented in Figures 4.2a and 4.4a). Task 1c allowed Howard to make the issue of systematic enumeration of all possible cases an explicit point for discussion in the class. Specifically, she asked students to compare Jim's list (line 28) with that of Harriet (line 29), thus eliciting from students two important points about following a systematic way to enumerate all possible cases in a finite set: first, this way helps ensure that the list does not include a case multiple times (line 30); second, it helps ensure also that no case is omitted from the list (line 31).

Jim's list in line 28 as part of his work on Task 1c, together with the comments that followed up on that list by Jim and others in the class, could be considered as offering a proof for the claim that there were six ways to make the number 10. This is because the argument underpinning this work had the following characteristics:

(1) It used true statements that were accepted by the class, notably the results of the six simple addition sentences in the list.

(2) It employed the valid mode of argumentation associated with the systematic enumeration of all possible cases in a finite set: This was achieved by starting with the smallest possible whole number as the first addend of the number sentence (i.e., $0 + 10 = 10$); moving on to the next whole number as the first added of the next number sentence; and proceeding like that up to the point when the next number sentence would have formed a commutative pair with a number sentence already included in the list.

(3) It was communicated appropriately using written language that the community could understand.

Task 1c′, which was an extension of Task 1c, offered the students an opportunity to apply or practice the mode of argumentation associated with enumerating systematically all possible cases in a finite set. A few students were still producing disorganized collections of number sentences (e.g., Figure 4.5a), with the arguments underpinning their work being considered to be *empirical*, as they offered inconclusive evidence that all possibilities were found. However, the majority of students were producing organized collections (e.g., Figure 4.5b, c).

In one of the whole class discussions around Task 1c′ there were a couple of instances when students announced that they had found a different number of ways than was actually possible given the conditions of the task: Pam said she had found 16 ways to make the number 10 (line 34), while Zenia said she had found five ways to make the number 14. These instances created a need for a proving activity to *refute* a claim. Pam's claim was refuted when she could not offer any other number sentences besides those that were in the organized list on the board. Pam admitted that she had not done her sentences "in order" (line 36), and so it is possible she had written down some sentences multiple times. Zenia's claim was refuted with an argument by *reductio ad absurdum* that was expressed implicitly by Howard (line 37) and that seemed to have convinced Zenia about the falsity of her claim (line 38). Specifically, Howard asked Zenia whether she thought there would be fewer ways to make the number 14 than there were ways to make the number 10 (line 37). With this question, Howard implied that a possible acceptance of Zenia's claim would contradict the premise, which Zenia and the rest of the class seemed to have tacitly accepted, that there would be at least as many ways to make a larger number (in this case 14) as there were ways to make a smaller number (in this case 10).

The work of the class on Task 1c′ concluded with students looking across their results on the number of ways to make different whole numbers from 10 to 20 and beginning to notice a pattern (lines 42–46). Howard's final comment during the lesson, "I wonder whether we could say how many ways there are for *any* number" (line 47), was essentially hinting at a new proving task that would involve making and *justifying* a generalization over an *infinite* number of cases. However, there was no time left during the lesson for the students to engage with this new task.

Concluding remarks

Episode A in Howard's class offered a vivid example of the rich and multifaceted mathematical activity that the purposeful design and implementation of a proving task with ambiguous conditions can help generate in an elementary classroom. The different conditions of the main task and of its variants or extension, together with the way in which Howard clarified these conditions gradually during the lesson and facilitated a smooth progression from one task to the next, helped engage the class in work that spanned sets of different cardinalities (both finite and infinite) and in different kinds of arguments to justify or refute claims about these cardinalities. In the context of this work, the students had an opportunity to learn or use important modes of argumentation, notably, describing procedures that could run *ad infinitum* as a way to justify the existence of an infinite number of possible cases, or enumerating systematically all possible cases in a finite set as a way to justify a claim about the set's cardinality.

But the mathematical activity of the class during the episode was not exclusively about proving. Rather, proving was the vehicle through which the class engaged meaningfully with other important mathematics too. Specifically, the various proving tasks with ambiguous conditions that were used by Howard in this episode offered students opportunities to also develop sensibility in mathematical language as this applies, for example, in one's efforts to figure out what is permissible by the conditions of a given task. Furthermore, the students had opportunities to solidify or deepen their knowledge of key arithmetical properties (e.g., ways of preserving a sum while changing the addends), concepts (e.g., commutative pairs of number sentences), and operations (e.g., addition involving a negative and a positive integer). The arithmetic work of the class, especially its part that concerned description of procedures for generating number sentences that could run *ad infinitum*, fostered the sort of reasoning about number and operations that underpins algebraic thinking (e.g., Carpenter et al., 2003; Kaput et al., 2007; Stacey, Chick, & Kendal, 2004).

Episode B[5]

Description of the Episode

This episode occurred on October 3 and describes part of the work of Ball's third-grade class on integer addition and subtraction, which began on September 26. While measured against the usual elementary school curriculum the topic of integer arithmetic might seem premature, Ball had good reasons for getting her class into this mathematical territory (her reasons are explained later). But in any case, the issues related to proving that are raised by the episode and are discussed in this chapter are not specific to the particular mathematical context (integer arithmetic) in which the work of the class was situated.

What had prompted the class to investigate aspects of integer arithmetic was the students' conception that "one can't take nine away from zero," which on September 25 made many students think that $300 - 190 = 290$. Ball noted in her journal that day: "[E]xpanding [students'] working domain for numbers seems a reasonable priority before either estimation competence (number sense) or precision with computation" (p. 26).

Ball wanted to make available to students a representational model to support their work. After considering different models—money (and debt), a frog on a number line, and game scoring—she decided to start with the "building model" in Figure 4.7. The building in the model consisted of 12 floors below ground, the ground floor (called the "0th floor"), 12 floors above ground, and a roof. Ball decided to use the circumflex (ˆ) above the numerals instead of the minus sign for negative integers. Her decision not to follow the standard mathematical notation was based on the belief that substituting the circumflex for the minus sign would help students focus "on the idea of a negative number as a *number*, not as an *operation* (i.e., subtraction) on a positive number" (Ball, 1993, p. 380; emphasis in original). For consistency, I use the same notation in my description and discussion of the episode.

The class used the building model to figure out answers to number sentences with integer addition and subtraction in which negative integers appeared only at the beginning of the sentences. To interpret and figure out answers to these number sentences, students imagined that each number sentence represented the trip of a person in the building. For example, the number sentence $\hat{5} + 2 - 1 = ?$ would be interpreted and solved as follows:

> The first term indicates the person's starting position. In this case, the person begins five floors below the ground floor. The addition operation indicates that the person has to go *up* the building. The second term indicates that the person has to travel two floors in this direction. The subtraction operation indicates that the person has to go *down* the building. The third term indicates that the person has to travel one floor in this direction. The person ends up four floors below the ground floor, so the answer to the number sentence is $\hat{4}$.

[5] In describing and discussing this episode I use with permission parts of Stylianides (2007a) (license number 3487821307724).

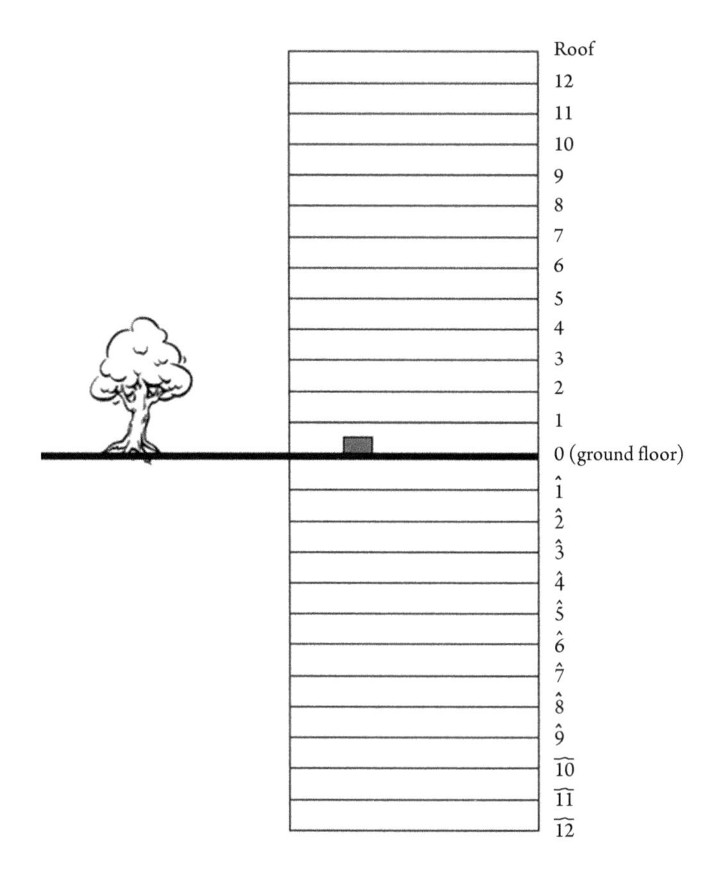

Figure 4.7 The building model.

On October 2 the students worked on finding different ways for a person to get to the second floor. The episode took place the next day, October 3, and it started with Ball giving the following task to the class:

1. *Task*: "How many ways are there for a person to get to the second floor? Prove your answer."

In her journal entry after the lesson (October 3, p. 37), Ball noted the following about the task:

2. The emphasis here, compared to yesterday, was on figuring out and justifying how many different ways there are for a person in the building to get to the second floor. I knew that those kids who only wrote two-addend number sentences (e.g., $\hat{4}+6=2$) would have 25 answers but those, like Lisa, who wrote multi-addend number sentences (e.g., $\hat{6}+10+3-2+1-4=2$) would have infinite solutions.

The students worked on the task individually or in small groups. At some point Ball called them back to the whole group to clarify the task:

3. *Ball*: [. . .] Maybe you finished finding all the ways, maybe you didn't. I want you to think right now about how *many* ways there are. Did you find them all? If you found them all, figure out how many there are and *prove* that's all there are. If you didn't find them all, write down what you think about how many there should be and *why* you think that.

The students continued their work on the task for another 10 minutes. Then Ball called them back to the whole group and asked them to share their work. Among several volunteers Ball chose Riba to share her work first. Riba went up to the board, stood by the building model, and said:

4. *Riba*: See, look [pause] aren't there 25 numbers [she pointed to the floors at the building model]? Then there have to be 25 'cause [pause] 25 answers, because you can't make more, because there are only 25.

Nathan pointed out that one would have to also count the roof to find 25 floors, but Riba said that there were 25 floors without considering the roof to be a floor. Ball asked Riba to show to Nathan why she thought there were 25 floors. Riba counted the floors one by one, beginning from floor $\hat{12}$ and ending at floor 12. Nathan admitted that Riba was right. Riba then offered to give a "different explanation," as she said, about why there were 25 ways for a person to get to the second floor:

5. *Riba*: This is twelve below zero [she pointed to the lowest floor]. If you write twelve below zero in your notebook [she writes $\hat{12}$ on the board], you would [pause] I'm saying, look, take twelve below zero. Then you take [she counted floors up, from floor $\hat{12}$ to floor 2] one, two, three, four, five, six, seven, eight, nine, ten, eleven, twelve, thirteen, fourteen! Plus fourteen equals two [she finished writing the following number sentence on the board: $\hat{12} + 14 = 2$].

Riba then noted that the same idea would apply for all 25 floors, not just floor $\hat{12}$. Ball asked Riba to write down some more number sentences and helped her record the sentence that corresponded to the trip beginning at floor $\hat{1}\hat{1}$: $\hat{1}\hat{1} + 13 = 2$. Riba continued from there.

6. *Riba*: And there could be another one and another one. You see, ten below zero plus [pause] this would have to be twelve equals two [she wrote $\hat{1}\hat{0} + 12 = 2$ on the board]. You have to keep on going like that, so you will finish at one below zero, and then go to [the] ground [floor] and you say: zero plus two equals two [she wrote $0 + 2 = 2$ on the board]. Like that. And then keep on going, keep on going, keep on going until this, and after you finish you go to twelve [she pointed to the top floor] and then you go this way and you get to two [she moved her hand downwards, from floor 12 to floor 2]. I'll show you this other way. Twelve [she wrote 12 on the board] there is [she counted floors starting from floor 11 and going down to floor 2] one, two, three, four, five, six, seven, eight, nine, ten! Then

it will be twelve take away ten equals two [she finished writing the following number sentence on the board: $12 - 10 = 2$]. I stopped at two, and that's why it equals two! [. . .]

7. *Ball*: So you are saying that there are 25 answers?
8. *Riba*: Yeah!

Ball invited comments from the rest of the class. Betsy asked for some clarification and Riba explained her thinking again. Ball then summarized Riba's argument:

9. *Ball*: [. . .] What she is saying is that you can start in every floor and then add up to two [she showed up the building to the second floor] or subtract to two [she showed down the building to the second floor]. And there are 25 floors, so she got 25 answers. That's all she is saying right now. Does that make sense or not?

The class seemed to have a good understanding of Riba's argument and Ball moved on, asking whether others found a different answer to the task. Several students said they found more than 25 ways. Jeannie said she found 26 ways while Lucy said she could not read the number of ways she thought there were because the number was "big." Ball decided to give the floor to Lucy. In her journal entry after the lesson that day (October 3, p. 38), Ball explained the rationale for her decision:

10. I deliberated for a moment about whether to have Jeannie write all her 26 ways up so we could see *how* she got more than Riba's 25. Since I knew she'd only written two-addend number sentences, I decided it would not be a fruitful use of time just to discover that she'd written one down more than once. Still, this is a dilemma, for will Jeannie understand that it *has* to be 25 if you only write two-addend sentences? Or will she not understand that, logically, there can only *be* 25? [emphasis in original]

When Lucy came up to the board, she said that she had written number sentences like $\hat{12} + 24 - 10 = 2$, and this was how she decided there was a "big" number of ways. Lisa was the next person to come up to the board. She said she had followed an approach similar to Lucy's, and she recorded the number sentence $\hat{11} + 11 + 2 = 2$.

Then Ball asked the class to think what the solutions of Lucy and Lisa had in common, and how this was different from what Riba had presented earlier. After eliciting some ideas from the students, Ball pointed out that in Riba's solution the person in the task went to the second floor in a single stop, whereas in the solutions of Lucy and Lisa the person made more than one stop en route to the second floor. Ball's journal entry after the lesson that day is illuminating of her thinking (October 3, p. 39):

11. I decided to push the point that Lucy and Lisa had both made the assumption that the person in the problem could make more than one "stop" en route to the second floor (e.g., $\hat{5} + 9 - 2 = 2$). This seemed, for the moment, the most important issue relative to the problem at hand: of how many solutions there are to the problem "how many ways are there to get to the second floor." *With each of the assumptions*

> *kids have made, the answer would be different* (25 vs. infinite) and I wanted the kids to see that, and to begin to develop a sense for the role of *assumptions.* [emphasis in original]

In trying to ensure that the students understood what Lucy and Lisa had done, Ball invited the class to describe a trip with multiple stops en route to the second floor. The students made up a trip that corresponded to the following number sentence: $6-1+7-8-2=2$. As soon as the class concluded the description of the trip, Lucy observed that the person could keep going up and down many times before finishing his trip at the second floor. This observation provoked Betsy's reaction. Specifically, Betsy objected to the way Lucy and Lisa solved the task by saying the following: "The person wants to go to the second floor—he doesn't want to go all over the building." Lisa defended hers and Lucy's approach by saying that, even if the person made multiple stops, he *would* eventually get to the second floor.

Ball asked the other students in the class what they wrote in their notebooks in solving the task. Four students said that they included solutions like those of Lucy and Lisa. The rest of them said that they included only two-addend number sentences like Riba had done. Ball said the following:

12. *Ball*: Neither one of these is right or wrong. Both are okay, but they are important. There is an important thing here for you to look at. Everyone look up here, please. Lucy and Lisa and Mei and, I don't remember who else [students mentioned Sean] and Sean made a different, what we call, *assumption.* I'm going to write the word on the board because it's an important word for us [she wrote the word "assumption" on the board]. That means that they thought something a little bit different than what Betsy's assumption was. Betsy's assumption was that the person wanted to go as fast as possible to the second floor. The problem doesn't say that, but it's okay that she thought that. That was her assumption. That means: that's what she thought. And then she did all her work on the problem because of that way of thinking. Right, Betsy? [Betsy affirmed.] So did some of the rest of you. Like Cassandra assumed the same thing. [...] She assumed that the person wanted to get there quickly. What was Lisa's assumption? What did Lisa think about the person? Betsy?
13. *Betsy*: That he wanted to take the whole day going from place to place. [Students laughed.]
14. *Ball*: Maybe he wanted to visit people, or wasn't in a rush. The problem doesn't say that, does it? [Students agreed that the problem did not specify that.] But it also doesn't say that he didn't. So they made two *different* assumptions. They had two different ideas. Lisa did her work after she made her assumption. You [she turned to Betsy] did your work after you made yours. And both are right, but they are different assumptions. If you make Betsy's assumption, if you assume that the person wants to get to the second floor as fast as possible, how many ways are there for the person to get to the second floor? [pause] If the only thing you are going to let the person do is start somewhere and go right to the second floor, in how many ways can the person do that? Chris?
15. *Chris*: Twenty-six.

16. *Ball*: Why 26?
17. *Chris*: Because he can also start from the roof.
18. *Ball*: Okay. He can start from the roof or every other floor and go to the second floor.

Ball remarked that the proof for 26 ways would be similar to the proof that Riba had presented earlier, but in this case one would also have to consider the roof as another possible starting point for the trips. She also pointed out that Chris made a different assumption. The class stated Chris' assumption: "that you can also count the roof [as a possible starting point for the trips]." Then Ball raised the question:

19. *Ball*: If you assume what Lisa and Lucy and Sean assumed, that the person wanted to travel around, how many ways are there for the person to do it?

Ball called on Ofala to answer this question because, as she noted in her journal entry after the lesson that day, she knew that Ofala had in her notebook an idea about "afinidy" (meaning to say "infinity"). However, Ofala did not seem to remember this idea.

Ball then called on Lucy who had said earlier that she could not read the number of ways she thought there were because the number was "big." Ball asked her to come up to the board to write this number. Lucy wrote on the board the number 8,000,000,000,000,000, 000,000,000, commenting further that she wanted to "add more zeroes" to the number but there was not enough space for those in her notebook.

20. *Ball*: What are you [she referred to Lucy] trying to say with this number?
21. *Lucy*: I'm trying to say that there is *a lot* of them [i.e., ways for the person to get to the second floor]!

Then Ball called on Jeannie, who said:

22. *Jeannie*: I think there are as much as you want because [pause] he can spend months going up and down if he wanted to! [Students laughed.]

The discussion continued for a few minutes and the lesson ended. Several students tried to explain how many ways there would be if one assumed that the person could travel around. Just before the end of the lesson Mei said: "The answer goes on for ever." However, there was not enough time left for her to explain what she meant by that.

Ball concluded her journal entry (October 3, p. 41) about the lesson that day with a reflection on students' emerging understanding of the notion of infinity:

23. I think people got the sense that there were a LOT of answers, given the second assumption [i.e., that the person could make multiple stops en route to the second floor]. But a LOT can mean 28 or 1,000 or 9,000,000,000,000 to an eight-year-old. And "endless" doesn't necessarily mean something different from a very big number—big numbers are themselves endless to these kids, I think. (This is a new twist on the idea of the confounding of "infinity" as a very big number.) [emphasis in original]

Discussion of the Episode

The proving task in which Ball engaged her students in this episode (line 1) was purposefully ambiguous, thus allowing students to formulate different legitimate assumptions about the task's conditions. In particular, there were two main (different) assumptions a student could make about the nature of the person's trips in the task: that the person would have to follow a direct route to the second floor or that the person could make multiple stops en route to the second floor. Depending on which assumption a student made (consciously or unconsciously), the student would engage in a proving activity that would have to consider either a *finite* or an *infinite* number of possible cases, with each activity leading to *justification* of a different claim: that there were 25 ways or infinitely many ways for the person to get to the second floor.

In implementing the task, Ball had two primary goals, both of which directly related to the fact that the conditions of the proving task were subject to different legitimate assumptions. First, she aimed to help her students develop arguments (and, if possible, proofs) based on each of the two main assumptions students could make about the conditions of the task (line 2). Second, she aimed to help her students understand that the apparent conflict between the different conclusions students could reach by following each of these assumptions (25 ways versus infinitely many ways) was due to the different assumptions they started from (line 11).

In what follows I discuss the work of the class on the proving task in the episode, which is summarized in Figure 4.8. I continue with a more focused discussion on the key role that the two main assumptions about the task's conditions played in that work.

The work of the class on the proving task

Riba's approach to the task was based on the assumption that the person in the task would go directly to the second floor, even though she did not seem to be conscious of the fact that she was making this assumption. Riba provided two related arguments about why, given her assumption, there were exactly 25 ways (lines 4–6).

Riba's first argument, also summarized by Ball (line 9), went as follows: "The person can begin from any floor and travel directly from there to the second floor; because there are 25 floors, there are 25 (different) ways for the person to get to the second floor." This argument was based on the statement that there were 25 floors in the building, which was not readily acceptable by all students in the class and thus could not be considered to belong to the community's set of accepted statements the given time. One of the students who did not accept this statement was Nathan who claimed that one would need to also count the roof to be able to say that there were 25 floors in the building. Riba refuted Nathan's claim by counting, one by one, all of the floors in the building. By establishing that there were indeed 25 floors in the building, Riba's argument qualified, given her assumption, as a proof in the particular context. This is because:

(1) It used statements that belonged to the classroom community's set of accepted statements, notably that there were 25 floors in the building.

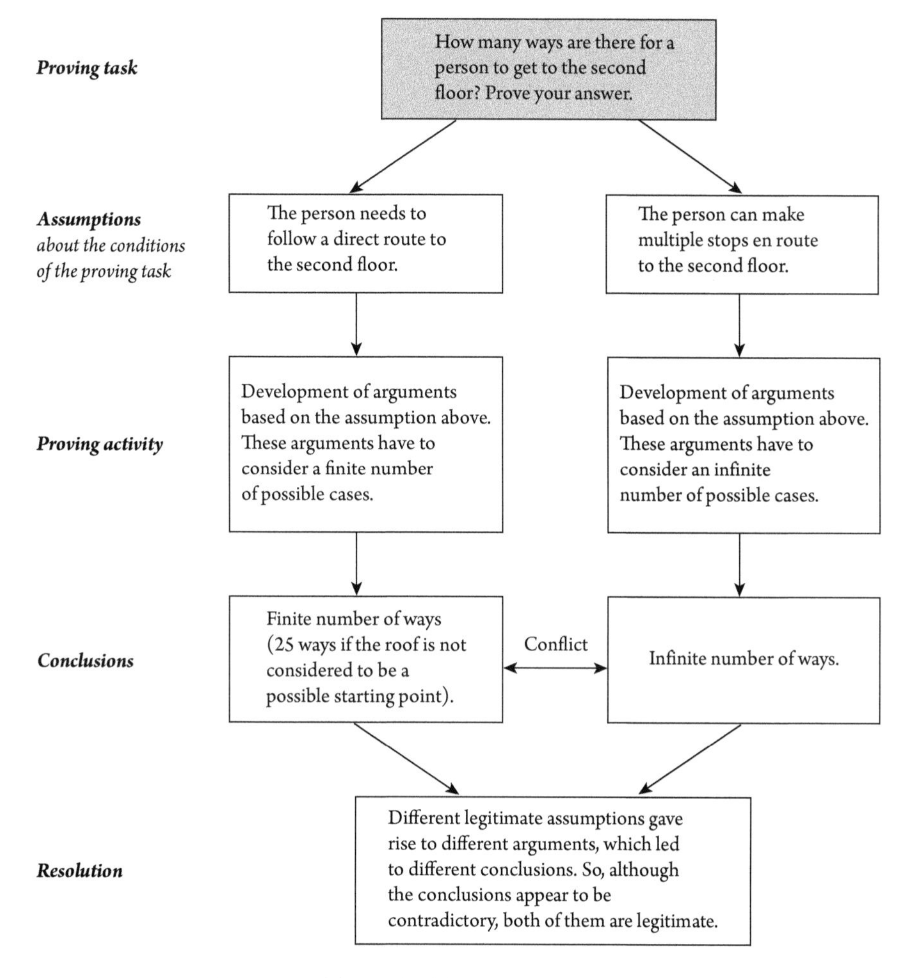

Figure 4.8 Summary of the work of the class on the proving task in Episode B.

(2) It employed the following mode of argumentation that was valid and, presumably, within the reach of the classroom community: each new direct trip to the second floor would have to start from a different floor in the building.

(3) It was represented appropriately using verbal language that was understandable to the students in the class.

Riba's second argument took the form of systematic enumeration of all the possible trips to the second floor, but instead of listing the number sentence corresponding to each trip she described in sufficient detail how one would go about generating those number sentences and gave representative examples on the board (lines 5 and 6). This was essentially outlining a different proof, which employed the valid mode of argumentation associated with systematic enumeration of all possible cases in a finite set.

After the presentation of Riba's arguments, Ball directed the students' attention to the other main assumption one could make about the task's conditions, namely, that the person could make multiple stops en route to the second floor. She called on Lucy and Lisa who wrote on the board multi-term number sentences to illustrate what made them think that the number of ways for the person to get to the second floor was "big."

The two different approaches to the task that were followed by different groups of students, and the apparent conflict between the conclusions that were derived from each of these approaches, generated controversy in the class. Some students (e.g., Betsy) thought that the "direct route" approach was the only legitimate approach. Other students (e.g., Lisa) thought that the "multiple stops" approach was also legitimate, for the person does eventually get to the second floor. This controversy raised naturally the need for the teacher to resolve the issue by highlighting to students the role of assumptions in what the students had been debating about. Ball explained to the students that each approach was based on a different assumption (lines 12 and 14). She called the students' attention to the term *assumptions* and she tried to help them understand the relation between different assumptions and different conclusions by asking them to consider what would be the answer to the task given each assumption (lines 14 and 19).

The students addressed easily the case in which the person in the task would follow a direct route to the second floor; they did so by referring to Riba's earlier arguments in the episode. However, the students faced difficulties in addressing the case in which the person in the task could make multiple stops en route to the second floor, presumably because the solution set in this case had an *infinite* cardinality. Although several students expressed ideas that approximated the notion of infinity, they seemed unable to clearly articulate or explain their thoughts (lines 20–22). For example, Jeannie claimed: "there are as much [ways] as you want" (line 22). She also tried to offer an argument for her claim by saying that the person in the task could "spend months going up and down if he wanted to" (line 22). This argument could potentially be the basis of a proof along the following lines: "The person can go up and down, visiting floors in the building before he ends up at the second floor, so that in each new trip the person makes a different number of stops than in any previous trip. Because the number of stops the person can make extends *ad infinitum*, the number of possible trips is infinite."[6] However, it is an empirical question whether third graders can formulate, or understand, an argument such as this one.

The role played by assumptions in the work of the class

The two main assumptions that students made about the conditions of the proving task played a twofold role in the work of the class in this episode. This role was facilitated by the

[6] For example, the person could make the trips according to the following procedure. For the first trip, the person goes up and down *once* between the third and fourth floors before traveling down to the second floor. For each new trip, the person goes up and down *one time more than in the previous trip* between the third and fourth floors before traveling down to the second floor.

design of the task and came to fruition with the way in which Ball implemented the task and orchestrated students' engagement with it.

The first role is summarized in Figure 4.8. This was as a means for engaging students, in the context of a single proving task, in two different kinds of proving activity: one that involves a *finite* number of cases and another that involves an *infinite* number of cases. Accordingly, the students had an opportunity to develop a sense of the fundamental mathematical idea that certain modes of argumentation (e.g., systematic enumeration of all possible cases in a finite set) may be more relevant to one kind of proving activity and less relevant to the other.

The second role that assumptions played was as a means for resolving a *problematic* situation, namely, the apparent conflict between the different conclusions reached by different groups of students for the number of possible trips to the second floor. I use the term "problematic" to describe a situation that both allows and encourages students "to problematize what they study, to define problems that elicit their curiosities and sense-making skills" (Hiebert et al., 1996, p. 12). In the case of this episode, the apparent conflict between the different conclusions reached by different groups of students both allowed and encouraged the class, led by the teacher, to examine the arguments that gave rise to these conclusions and search for a resolution. The teacher guided her students to see that the situation could be resolved by reference to the different (legitimate) assumptions that constituted the basis of these arguments and, thus, of their conclusions, too. As a result, the students had an opportunity to develop a sense of the fundamental mathematical idea that the truth of conclusions depends on the assumptions that support them. According to Fawcett (1938), proving offers a natural context for children to understand the relation between conclusions and their underlying assumptions, thereby fostering children's ability for *reflective thinking* (Dewey, 1910/1997), that is, "[a]ctive, persistent, and careful consideration of any belief or supposed form of knowledge in the light of the grounds that support it, and the further conclusions to which it tends" (p. 6, original in italics).

Concluding remarks

Ball engaged her class in a proving task whose conditions were deliberately ambiguous and thus subject to different legitimate assumptions by students. This episode illustrated the different kinds of proving activity that such a task, when carefully designed and purposefully implemented, can help generate in the classroom. In addition, it described the first time when the students in Ball's class were introduced to the role of assumptions in proving, thus offering a vivid image of what it might look like when the different assumptions about a task's conditions become themselves an object of study in an elementary classroom. This complements other similar images described in the literature, but these involved secondary students and were situated in the context of upper-level mathematics (Fawcett, 1938; Jahnke & Wambach, 2013).

Interestingly, the work of the class in this episode, scaffolded by the teacher, has some similarities to the work of mathematicians in the discipline. Take, for example, mathematicians' work related to the development of Euclidean and non-Euclidean geometries (this is presented in detail in Fawcett, 1938). Similar to the adoption of different sets of axioms by Euclid, Riemann, and Lobatchewsky, different groups of students in the class made (unconsciously) different assumptions about the task's conditions. Also, similar to the attempts of Euclid, Riemann, and Lobatchewsky to build self-consistent geometrical theories based on their different sets of axioms, the class attempted to construct valid arguments based on their beginning assumptions. Furthermore, similar to the idea that the different sets of axioms of Euclidean, Riemannian, and Lobatchewskian geometries account, for example, for their "contradictory" conclusions for the sum of the interior angles of a triangle (equal to, more than, and less than 180°, respectively), the class resolved the issue of "contradictory" conclusions for the given task by reference to the different assumptions that supported the respective arguments. Finally, similar to the fact that mathematicians' development of different geometrical theories based on different sets of axioms promoted the field's understanding of geometrical space, the students in the class were offered opportunities to enhance their understanding of proving and develop a sense of the role of assumptions in this activity.

Of course, no one denies that children's work differs from of that of mathematicians in many respects (e.g., Hiebert et al., 1996, p. 19). Nevertheless, and given the similarities described earlier, it may be useful to view students' work with assumptions in Episode B as an *authentic* (Lampert, 1992), albeit *rudimentary* (Bruner, 1960), version of mathematicians' work with assumptions in the discipline. In particular, one may describe as follows the first and last points of a possible *learning trajectory* (e.g., Maloney, Confrey, & Nguyen, 2014; Simon, 1995) regarding students' engagement with assumptions in school mathematics: firstly, develop an understanding of the role played by assumptions in the "local" setting of proving tasks with ambiguous conditions (as in Episode B); lastly, develop an understanding of the role played by assumptions in the more "global" setting of mathematical theories that are grounded on different sets of axioms (as in the episodes with secondary students described in Fawcett, 1938; Jahnke & Wambach, 2013).

General Discussion

Episodes A and B offered complementary views on what it might mean or look like when elementary teachers engage their students with proving tasks that have ambiguous conditions. In this final section I discuss some general issues about the relationship between proving tasks and proving activity, in the particular context of proving tasks with ambiguous conditions, and about the role of the teacher while implementing this kind of task in the classroom.

The Relationship between Proving Tasks and Proving Activity

The role of assumptions

Episodes A and B suggest the role that assumptions play, or can play, as mediators of the relationship between proving tasks with ambiguous conditions and the proving activity that these tasks can generate in the classroom. This role can be explicit and even made transparent to the students, as in Episode B, or it can be implicit and stay in the background while still exercising an important influence on students' work, as in Episode A.

In Episode A, several students identified ambiguities in the conditions of the main proving task and asked the teacher questions so as to clarify the task's conditions. Howard did not clarify the conditions of the task in their totality right away. Rather, she clarified them gradually during the lesson, with each point of clarification that she offered producing a variant of the main proving task that engaged the class in important mathematical activity. Thus, while the ambiguous conditions of the main proving task in Episode A were, or could have been, subject to different legitimate assumptions by students, we did not see students in the episode actively pursuing different assumptions. Rather, the different assumptions that students could have made or pursued were embedded in the variants of the main task with which Howard engaged the class during the episode.

In Episode B, on the other hand, different groups of students pursued independently different assumptions about the ambiguous conditions of the given proving task, and they constructed different arguments based on these assumptions, reaching conclusions that appeared to contradict each other. Ball, who had anticipated these different student interpretations of the task's conditions, exploited the problematic situation that emerged in the class from students' "contradictory" conclusions to point out to them the role that different assumptions played in that work. Further, she made the notion of "assumptions" an object of study during the lesson.

Thus, the different assumptions that students can legitimately make about the ambiguous conditions of a proving task can mediate the relationship between proving tasks and proving activity in at least two ways: explicitly, serving as the factors, originally "hidden," that can explain why apparently contradictory conclusions that had derived from different student arguments can all be legitimate; and implicitly, underpinning from behind the scenes variants of the proving task as its conditions become clarified by the teacher. The proving task itself, in combination with the teacher's goals for using the task in the classroom, can determine which of these or other roles assumptions may play.

The nexus between assumptions and definitions

The two episodes also raise a broader issue about the nexus between the notions of "assumptions" and "definitions," in the context of proving tasks with ambiguous conditions and in relation to the proving activity that these tasks can generate in the classroom. In particular, one can observe that clear definitions of key terms in the statements of the proving tasks in the two episodes would have ruled out the ambiguity in the tasks' conditions and thus would have prevented different assumptions about those conditions. Specifically, clear definitions of what counted as a "way" to make the number 10 in Episode A, or what counted as a "way" for the person to get to the second floor in Episode B, would have clarified, for example, whether commutative pairs of number sentences represented different ways in Episode A, or whether the person had to follow a direct route to the second floor in Episode B. If such definitions had been offered, the proving activity of the two classes would have been fundamentally different from that in the episodes.

Let me consider another example of a task to probe a bit further into the nexus between assumptions and definitions. Following an investigation involving quadrilaterals, the task asks students to explore the truth or falsity of the following statement: "Every rectangle is a trapezoid." Depending on what one takes the definition of a trapezoid to be—"a quadrilateral with *at least one* pair of parallel sides" or "a quadrilateral with *exactly one* pair of parallel sides"—the statement is true or false, respectively. In each case, the statement can be justified or refuted using different modes of argumentation.

This example shows that one possible form assumptions can take in the context of proving relates to the choice of definition. Zaslavsky (2005) noted that the choice of definition need not be connected to correctness, but rather "[i]t could be related to personal preferences, beliefs, values or the theoretical framework or context to which one refers" (p. 301). Indeed, the choice of the definition of a trapezoid as "a quadrilateral with *at least one* pair of parallel sides," which would allow a rectangle to be considered a kind of trapezoid, could reflect a value of the fundamental mathematical idea of *generalization* (e.g., Kitcher, 1984) or a value of *hierarchical classifications* of concepts (e.g., de Villiers, 1994). The example suggests further that a classroom community that shared a clear definition of a trapezoid would most likely not engage in discussions about the role of assumptions in justifying or refuting the particular statement, because the shared definition would have ruled out the ambiguity of the situation. This implies that definitions can influence the interpretation of mathematics tasks and raises the issue of how teachers can manage the tension between clear definitions and ambiguous task conditions that can support different assumptions.

The Role of the Teacher

A teacher who uses a proving task with ambiguous conditions may be faced with a multitude of options, and thus with some difficult decisions to make, about which assumptions or proving activity to pursue, or about when to use clear definitions to rule out the ambiguity in a task's conditions. In what follows I use the context of Episodes A and B to discuss factors that may inform a teacher's decisions in this pedagogical space.

A major factor that teachers can consider in making decisions in this pedagogical space is whether engaging the class in proving activity based on a particular assumption would leave behind an important *learning residue* (Davis, 1992), a term used to describe the student understandings that remain after an activity is over. The size of this residue would depend, then, on what other proving activity the class had already engaged in and the extent of overlap between the learning affordances of these activities. I use two examples, one from each episode, to illustrate these points.

In Episode B, besides the two main assumptions described in Figure 4.8 about the nature of the person's trip, there were some other points in the task's conditions that, in the eyes of a few students, were also subject to different assumptions. Specifically, although almost all of the students in the class assumed that the 25 floors of the building were the only possible starting points for the person's trips, Chris assumed that the roof of the building was another possible starting point (lines 14–18). Ball decided not to ask the students to explore the implications of different assumptions about the possible starting points for the person's trips, for, as she remarked during the lesson, the proof for 26 ways (based on the alternative assumption proposed by Chris) would be very similar to the proof for 25 ways that had already been presented by Riba. Thus, if Ball had asked the students to pursue Chris' assumption and construct an argument based on it, this work would have not left behind any important learning residue.

In Episode A, the assumptions that were embedded in the statements of Tasks 1, 1a, and 1b, as well as the proving activity that was generated by these tasks (Figure 4.6), might appear at first sight to be very similar to each other, thus questioning whether Task 1a, and even more so Task 1b, left behind any important learning residue. However, this remark is challenged by a more careful examination of the learning affordances of these tasks and the mathematical work that the tasks actually generated in the classroom. Indeed, as I discussed earlier, each of these tasks engaged the class in related but distinctively important mathematical work that extended beyond the area of proving. For example, each new task engaged students in describing a new procedure for generating number sentences for 10 that could run *ad infinitum*, imposing on their work constraints that pushed them to think deeply about different key arithmetical properties, concepts, and operations.

A related factor that teachers should consider, especially when managing the tension between clear definitions and ambiguous task conditions that can support different assumptions, is the specific learning goal they are trying to accomplish. If teachers aim to offer their students opportunities to develop a sense of the role of assumptions in proving (similar to what happened in Episode B), or opportunities to develop sensibility in mathematical language and experience with clarifying the conditions of a task (similar to what happened in Episode A), then unclear definitions and ambiguous task conditions may be appropriate. If, however, teachers are more concerned with introducing their students to a new mode of argumentation or proof method and less with increasing their understanding of the elements that constitute the foundation of an argument (assumptions, axioms, etc.), then clear definitions and unambiguous task conditions may be appropriate.

But there may also be situations where teachers aim to promote a combination of learning goals, such as an understanding of the role of assumptions in resolving "conflicting" conclusions in proving tasks with ambiguous conditions *and* an appreciation of the role of clear definitions in ensuring unambiguous interpretation of a task's conditions. An instructional sequence that could promote this expanded set of goals could include the following three phases: (1) the teacher poses to students a proving task with ambiguous conditions that allows students to generate "conflicting" conclusions based on different assumptions; (2) the teacher helps the students make their assumptions explicit and resolve the issue of "conflicting" conclusions by reference to their different assumptions; and (3) the teacher asks students to think what kind of definitions or clarifications would have helped rule out the ambiguity in the task's conditions so that everybody would have worked on the "same" task. Episode B illustrated the first two phases of this instructional sequence, while Episode A described a situation that approximated the third phase.

Finally, there may be situations where teachers do not realize in advance that a particular proving task they give to their students has ambiguous conditions. Even though a teacher may think that the task's conditions are clearly defined, it is important that the teacher is able to recognize in the course of teaching that some students' approaches to the task that appear to be "faulty" may be in fact mathematically sound and based on an unforeseen set of legitimate assumptions or definitions. The in-the-moment diagnosis of the deeper underpinnings of students' work is a crucial factor for an informed teacher response to students' work, but it nevertheless places heavy demands on the teacher's knowledge (e.g., Rowland et al., 2015; Stein et al., 2008; Stylianides & Stylianides, 2014b).

5

Proving Tasks Involving a Single Case

In this chapter I examine proving tasks involving a single case, with an emphasis on the proving activity that this kind of task can help generate in the classroom and on the role of the teacher while implementing the tasks. Specifically, I focus on proving tasks that are placed in the context of a single calculation involving a basic integer operation, such as addition of two multi-digit whole numbers. While this specific category of tasks does not cover the full range of proving tasks involving a single case, it nevertheless relates to a fairly large part of the mathematical work in elementary schools. Calculation problems are not normally associated with proving, and of course I would not consider *any* calculation problem to be a proving task. Drawing on the view of proving tasks as a special category of mathematically rich and cognitively demanding tasks that I discussed in Chapters 1 and 2, a calculation problem that would qualify as a proving task would have to entail, in addition to an expectation to justify or refute a statement, an element of mathematical challenge for the particular group of students to which it is addressed. The element of challenge in solving a calculation problem that meets the standard of a proving task should definitely surpass the straightforward application of a known calculation algorithm or procedure.

As in Chapter 4, I situate my discussion in the context of two classroom episodes: the first (Episode C) comes from Howard's Year 4 class in England (8–9-year-olds); the second (Episode D) comes from Ball's third-grade class in the United States (again 8–9-year-olds). These episodes have been selected for their illustrative power and for the complementary (notably contrasting) issues they raise about what it might mean or look like when elementary teachers use in their classes this particular kind of proving task. I describe and discuss each episode separately, and I conclude with a more general discussion.

Episode C

Description of the Episode

This episode took place in Howard's Year 4 class and describes part of a lesson that Howard taught on July 7. The lesson started with the students asked to do individually the following calculations in their notebooks:

1. *List of calculations*:
   a) 206×4
   b) 117×5
   c) 109×6
   d) 305×9
   e) one of your own . . .

The students used primarily the standard written method (calculation algorithm) taught in English elementary schools and presented their work vertically, as in the following example.

2. *Sample student work for the first calculation*:

$$\begin{array}{r} 206 \\ \times\ 4 \\ \hline 824 \\ \hline {\scriptstyle 2} \end{array}$$

In the whole class discussion that followed, Howard asked the students to describe their work, which they did by reciting the steps of the standard method as applied in the particular calculation. She also asked students about the meaning of different digits showing up in their written work, which they explained competently, demonstrating good understanding of place value. For example, they explained that the "little 2" that appears in the last row of the sample student work (line 2) stood for 20. This introductory part of the lesson finished with Howard asking students to say how many out of the first four calculations they had solved correctly. Most of the students said they solved all of them correctly, though some students said they made a few mistakes.

The main part of the lesson was devoted to solving the following calculation problem, which I consider to be a proving task:

3. *Missing Digits Calculation Problem*: "Find the missing digits to make the following addition (Figure 5.1) correct. Prove your answer."

This task, without the expectation for a proof, derived from a set of sample assessment materials that were released by the Standards and Testing Agency in the UK (Standards & Testing Agency, 2014, p. 7). Howard offered students the option to work together in pairs or

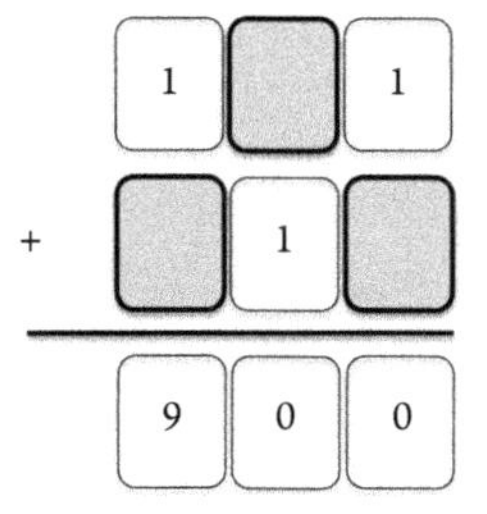

Figure 5.1 The calculation in the Missing Digits Calculation Problem.

larger groups and to use different resources as they saw fit (paper and pencil, base-10 blocks, counters, etc.).

During the small-group work, I had the opportunity to talk with two pairs of students. The first pair, Andy and Keith, said that originally they had made an error in the hundreds digit of the second addend (Figure 5.2a), which they then identified and corrected (Figure 5.2b).

When I asked them to explain their answer to me they responded in a way that can be paraphrased as follows:

4. *Andy and Keith's explanation* (Figure 5.2b): In the units column, we need a number which when added to 1 makes 10; this number is 9. We carry the 1 to the next column [pointing at the "little 1" written at the bottom of the tens column]. In the tens column, we need a number which when added to the two 1s makes 10; this number is 8. We carry the 1 to the next column [pointing at the little 1 written at the bottom of the hundreds column]. Finally we need a number which when added to the two 1s makes 9; this number is 7.

I also asked them to explain how they figured out their original error in the hundreds digits of the second addend (Figure 5.2a). They said that, after their first attempt to solve the problem, they tried to verify their answer and found out that the sum of 181 and 819 was 1000, instead of 900 as required. Then they spotted the error in the last step of the procedure

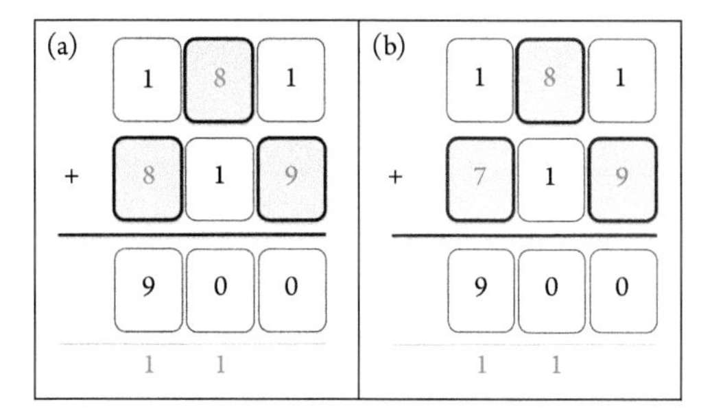

Figure 5.2 The original work (a) and the corrected work (b) of Andy and Keith to the Missing Digits Calculation Problem.

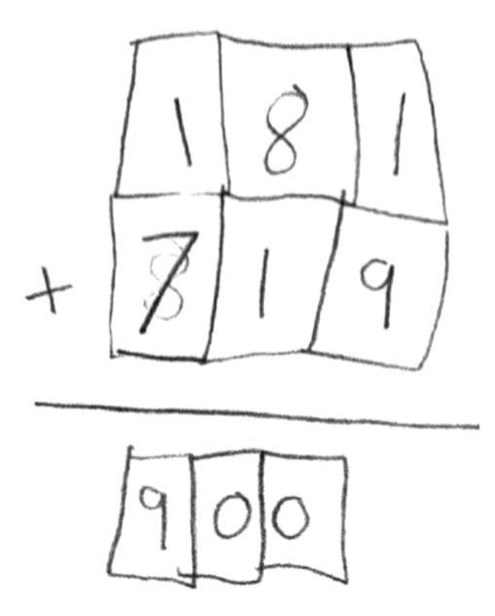

Figure 5.3 The corrected work of Ella to the Missing Digits Calculation Problem indicating the original error in the hundreds column.

involving the hundreds column. Several other students in the class, such as Ella (Figure 5.3), had originally made the same error, as evidenced by the corrections on their papers.

The second pair of students with whom I talked during small-group work, Simon and Cameron, offered essentially the same explanation as Andy and Keith did (line 4), but they presented their answer using counters. Figure 5.4 shows part of Simon's work, which included a key with the values of counters of different colors and sizes. In the places of the

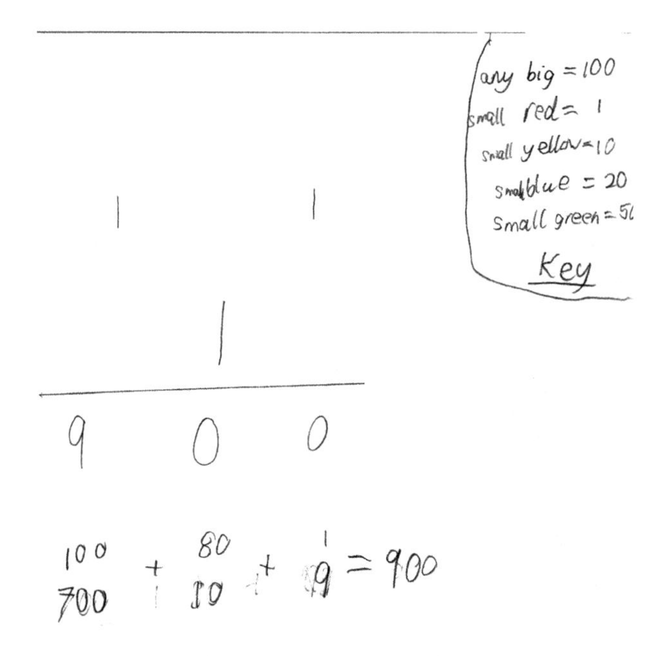

Figure 5.4 Part of Simon's work to the Missing Digits Calculation Problem using counters.

missing digits, Simon and Cameron had placed counters that amounted to the correct place value of the corresponding digits as summarized at the bottom of Simon's paper. For example, in the place of the units digit of the second addend, Simon and Cameron had placed nine small red counters.

The only notable difference between the explanations of the two pairs of students was that Simon and Cameron used the terminology of "jigsaw numbers" to describe to me number bonds for specific numbers. For example, they described 1 and 9 as jigsaw numbers to 10, and 80 and 10 as jigsaw numbers to 90. Howard explained to me after the lesson that the terminology of jigsaw numbers helped to make number bonds more accessible to children and memorable for them. Jigsaw numbers featured prominently in the whole class discussion that followed the work of the students in small groups.

5. *Howard*: What do you have to do [to solve the problem]?
6. *Gina*: Find the missing digits.
7. *Howard*: Where should people start, and why? [pause] Alex?
8. *Alex*: [thinks]
9. *Howard*: What did you do when you started?
10. *Alex*: I tried to find the jigsaw numbers.
11. *Howard*: What are jigsaw numbers?

After some discussion of jigsaw numbers, Howard directed the attention of the class back to the problem:

12. *Howard*: What are these? [She pointed to the two addends in the calculation.]
13. *Alex*: Jigsaw numbers to 900.
14. *Emma*: Let's start with the units.
15. *Howard*: What jigsaw numbers are we looking for for this missing digit? [She pointed to the missing units digit of the second addend.]
16. *Emma*: Jigsaw numbers to 10.
17. *Howard*: Why are you saying ten if this [the units digit of the sum] is zero?
18. *Emma*: You couldn't get a zero there [the units digit of the sum] with the 1 [the units digit of the first addend] and the missing number added together.
19. *Howard*: What's the missing digit then?
20. *Emma*: Nine.

Howard asked the students to continue their work on the Missing Digits Calculation Problem or, if they were done with it, to start thinking about an extended version of the problem involving addition of two four-digit numbers (Figure 5.5).

In the whole class discussion that followed, Howard focused students' attention on Ronnie's solution for the original problem, operating on the assumption that it had already been established in the class that the missing units digit of the second addend was 9, as proposed earlier by Emma (line 20).

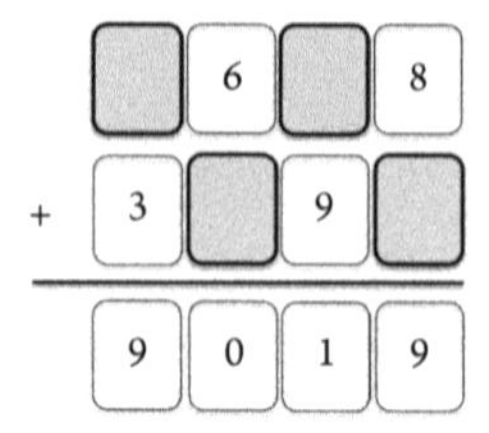

Figure 5.5 The calculation in the Extended Missing Digits Calculation Problem.

21. *Howard*: I'd like us to talk about Ronnie's solution. He looked at these four digits [see Figure 5.6] and wanted to make 100 with these two numbers [she pointed at the two-digit numbers formed by each pair of consecutive digits]. Why did he want to do that?

A couple of students explained that for the sum (i.e., 900) to have zeroes in its units and tens columns, the two two-digit numbers pointed at by Howard had to add up either to zero (which was impossible given that there were already several non-zero digits) or to 100. The class then agreed that the missing tens digit of the first addend had to be 8, and Howard continued her commentary on Ronnie's solution.

22. *Howard*: Then he [Ronnie] thought what [the digit in] the hundreds column [of the second addend] had to be to make this 900 [she pointed at the hundreds digit of the sum].
23. *Pam*: It has to be seven.
24. *Howard*: Why?
25. *Pam:* We already have 100 in the hundreds column from the first number [the first addend] and another 100 from the first two columns [the units and tens columns].

The class moved on to work further on the Extended Missing Digits Calculation Problem (Figure 5.5). Following work in small groups, the lesson finished with a whole class discussion focusing on parts of that problem.

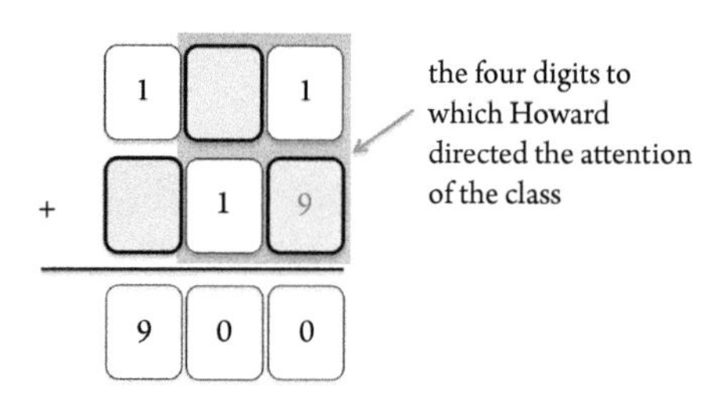

Figure 5.6 The four digits to which Howard directed the attention of the class during the discussion of Ronnie's solution.

Discussion of the Episode

The proving task with which Howard engaged her students in this episode (line 3) was about finding the missing digits in a *single* calculation problem (a sum of two three-digit numbers) and *justifying* that this was indeed the correct choice of digits. The extended version of the problem involving addition of two four-digit numbers (Figure 5.5) had the same general characteristics as the original proving task. Contrary to the Missing Digits Calculation Problem and its extended version, the calculation problems done by the class during the introductory part of the lesson (line 1) did not entail an element of challenge for students and did not comply with the general view of proving tasks as a special category of mathematically rich and cognitively demanding tasks that I discussed in Chapters 1 and 2. Indeed, students' engagement with these introductory problems involved a straightforward application of known procedures, with most students deriving the correct answers by drawing competently on their understanding of different aspects of place value (Thompson, 2003) and their mastery of the standard written method for addition.

The Missing Digits Calculation Problem engaged the class in a proving activity that led many students to the correct choice of digits and to valid arguments for their choice that met the standard of proof (according to the definition discussed in Chapter 2). Let me consider Andy and Keith's argument (line 4):

(1) It used true statements that were readily accepted by the classroom community, notably meanings related to different aspects of place value, the steps of the standard written method for addition, and simple addition facts (e.g., number bonds for 10).

(2) It employed the valid mode of argumentation associated with the correct application of the steps of the written method, but this was done flexibly and in a way that was new to the students to account for the fact that, contrary to students' past experiences, the missing digits were distributed over the two addends rather than belonging exclusively to the sum.

(3) It was represented appropriately in a way that was understandable to the classroom community, combining the students' written work when applying the method and their verbal explanation of the reasoning underpinning that work.

Simon and Cameron's argument was similar to that of Andy and Keith, with the only difference being that the representation of Simon and Cameron's argument also used a physical apparatus (i.e., counters of different colors and sizes) and the terminology of jigsaw numbers in describing number bonds for specific numbers. Furthermore, the whole class discussion that was orchestrated by Howard involved a number of students (lines 5–20) and helped unpick important elements of essentially the same argument, using again the terminology of jigsaw numbers. Howard's probing questions (e.g., line 17) helped students think more deeply about, and articulate more clearly, the place value ideas underpinning the various elements of the argument.

While most of the students in the class attempted to develop an argument based on flexible application of the steps of the standard written method for addition, a few others like

Ronnie approached the problem in an ad hoc way by taking advantage of the fact that the sum had zeroes in the units and tens columns. The argument outlined in the whole class discussion of Ronnie's solution (lines 21–25) could form the basis of another proof for the answer to the task. Yet another proof could be based on the observation (requiring justification) that the Missing Digits Calculation Problem is mathematically equivalent to a problem having the same brief but involving a different calculation, namely, the calculation presented in Figure 5.7. The re-formulated version of the problem could be solved by a simple subtraction. In a discussion I had with Howard at the end of the lesson, I mentioned to her the possibility of re-formulating the problem in this way. She said she liked the idea and she thought that the mathematical equivalence between the two versions of the problem could be accessible to her students.[1]

The proving activity of the class also involved the formulation of some subsidiary arguments that students used to *refute* some false statements connected to errors they had made in solving the Missing Digits Calculation Problem. An example of this is found in Andy and Keith's explanation of their error in the hundreds digit of the second addend (Figure 5.2a). Their explanation was essentially an argument by *reductio ad absurdum*: they explained that acceptance of their original answer to the problem led to a contradiction (when they added up the two numbers, they got a different sum from the sum specified in the problem). The emerging contradiction made these students search for and find the error in their work. Another form of refutation of their original answer could simply be the direct discovery of the error.

To conclude, the proving activity in which Howard engaged her class in this episode took place in the context of a task involving a single calculation and was intertwined with activity in the domain of elementary arithmetic. As the students were trying to find the missing digits in the given calculation and justify their choices of those digits, they were drawing upon their existing knowledge of place value and calculation methods. In fact the novelty of the particular task, with the missing digits being distributed over the two addends rather than belonging exclusively to the sum as in past calculations, afforded students an opportunity to

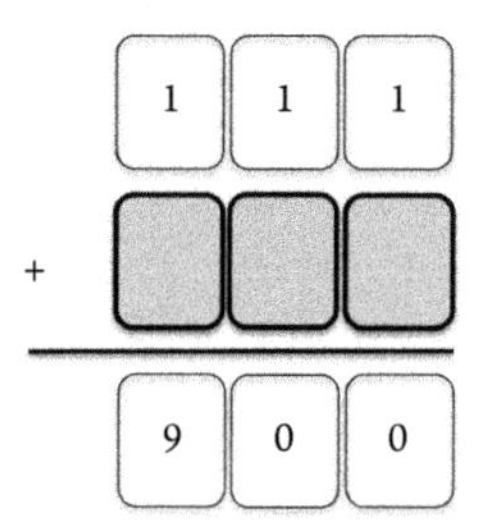

Figure 5.7 The calculation in a mathematically equivalent formulation of the Missing Digits Calculation Problem.

[1] Another possible proof for the Missing Digits Calculation Problem could use the mode of argumentation associated with systematic consideration of all the possible sets of three digits that could satisfy the given sum. However, this would be a tedious proof, and indeed none of the students in Howard's class attempted to produce it.

deepen their existing knowledge of place value and calculation methods. The teacher played a key role in asking students probing questions that encouraged them to think more deeply about the arithmetical concepts underpinning their arguments, thus supporting students' broader engagement in mathematics as a sense-making activity.

Episode D[2]

Description of the Episode

Similar to Episode B in Chapter 4, this episode describes part of the work of Ball's third-grade class on integer addition and subtraction, which began on September 26. In Chapter 4 I explained what had prompted the class to investigate aspects of integer arithmetic, Ball's considerations in choosing the "building model" (Figure 4.7) to support her students' work, and her decision to use the circumflex ($\hat{}$) above the numerals instead of the minus sign of negative integers. Here I reiterate an important point I made in Chapter 4: Although measured against the usual elementary school curriculum the topic of integer addition and subtraction might seem premature, Ball had good reasons (explained in Chapter 4) for getting her class into this mathematical territory. But in any case, the issues related to proving that are raised by the episode, and which I discuss in this chapter, are not specific to the particular mathematical context (integer arithmetic) where the work of the class was situated.

In Chapter 4 I also described few classroom events leading to Episode B on October 3. As a background for my discussion in this chapter, I describe briefly a couple of classroom events that happened after October 3, leading to Episode D on October 12.

On October 4 the students worked on Nathan's conjecture that "any number below zero plus that same number above zero equals zero," and they provided different arguments for it. One of these arguments, proposed by Sheena, was formulated in the particular case $\hat{12} + 12 = 0$ and went as follows: "If you start 12 below zero [in the building model], you will have to go up that same number [of floors] to get to zero, because zero is like in the middle of the numbers." Students ultimately began to make sense of number sentences of the form $\hat{n} + n = 0$, where n is a positive integer. One of the number sentences they had no trouble making sense of was $\hat{6} + 6 = 0$.

A few days later, on October 10, Lucy provided the following argument in support of the claim that the double of $\hat{8}$ is $\hat{16}$. First, she interpreted "doubling $\hat{8}$" to mean "$\hat{8}$ and another $\hat{8}$," i.e. $\hat{8} + \hat{8}$, and she reasoned that, since $\hat{8} + 8$ equaled zero, $\hat{8} + \hat{8}$ could not also be zero, operating on the assumption that the difference between the second addends of these two number sentences necessitated different sums. Then, she interpreted $\hat{8} + \hat{8}$ to mean "two eights below zero"; she extended the building model down to floor $\hat{16}$; and she modeled the addition sentence by starting on floor $\hat{8}$ and moving eight floors down, thus ending up on floor $\hat{16}$. This was the first time that a student in the class began to articulate how the class might deal with addition of a negative integer.

[2] In describing and discussing this episode I draw on parts of my unpublished doctoral dissertation (Stylianides, 2005).

On October 12, the day of Episode D, the lesson began with Ball asking the class to work on six problems, three of which were the following:

1. *Calculation Problems*: "What are the answers to these problems?

   a) $\hat{6}+\hat{6}=?$

   b) $\hat{6}+6=?$

   c) $6+\hat{6}=?$

   Justify your answers."

Ball explained in her journal entry later that day (October 12, p. 59) her reasoning for choosing these problems:

2. I selected these [problems] deliberately to provoke/elicit their recognition of relationships among each group—i.e., $\hat{6}+6$ was a problem they'd mostly become comfortable on the building [see description of classroom event on October 4] and $\hat{6}+\hat{6}$ was like the problem Lucy had reasoned on Tuesday in trying to figure out what number was the double of $\hat{8}$ (she'd argued that the double of $\hat{8}$ was $\hat{16}$ because it was like "*two* eights below zero") [see description of classroom event on October 10]. So, then, if $\hat{6}+\hat{6}$ was $\hat{12}$ (I expected that quite a few might make this connection) then $6+\hat{6}$ might follow similar logic. Another possibility which I keep expecting but which has *not* yet emerged is commutativity (i.e., $\hat{6}+6=0$, then $6+\hat{6}$ must also be zero). [emphasis in original]

After the students had worked on the problems individually and in their small groups, the class began discussion in the whole group. The discussion of the problem $6+\hat{6}=?$ evolved to what the teacher characterized in her journal entry at the end of the day as a "scientific debate": The students proposed and debated over multiple answers while only one of them was actually right; they persisted in questioning and challenging one another's claims; and they tried to resolve their disagreements publicly in a sensible way. The following segment is a small part of this lively debate, which lasted almost 30 minutes:

3. *Sean*: I got six [as the answer to $6+\hat{6}=?$].
4. *Betsy*: Instead of Sean's [answer], I got zero.
5. *Ball*: You'd like to put zero here for six plus minus six?
6. *Betsy*: Do you want to see how I do it?
7. *Ball*: Okay.
8. *Other students*: Yeah!
9. *Betsy*: Here. You're here [she shows the sixth floor on the building model], but you can't go up to twelve, because that's six plus *six*. So I say it's just the opposite. It's just six minus six.
10. *Sean*: But it says plus, not minus!
11. *Betsy*: But you're minusing.

12. *Riba*: Where'd you get the minus?
13. *Sean*: You should just leave it alone. You can't *add* six below zero, so you just leave it. Just say "goodbye" and leave it alone and it is still just six.
14. *Mei*: But this six below zero would just disappear into thin air?!
15. *Sean*: I know. It would just disappear because it *wouldn't* be able to do anything. It just stays the same, it stays on the same number. Nothing is happening.
16. *Betsy*: But, Sean, what would you do with this six below zero then?
17. *Sean*: You just say "goodbye" and leave it alone.
18. *Riba*: You *can't* do that. It's a *number*.
19. *Sean*: I know, but it's not going down. It's going up because it says plus.
20. *Mei*: I think I disagree with Betsy and Sean because I came up with the answer nine.

The discussion went on for a bit longer. Here is what Ball noted about the discussion in her journal entry at the end of the day (October 12, p. 63):

21. The class seemed exceptionally engaged in this argument [i.e., what is the answer to $6 + \hat{6} = ?$ and why]. I think everyone except Devin and Maria talked and perhaps all of those but Mark volunteered. They were arguing directly with one another, little through me. Striking also was that no one asks me what the answer is.

The lesson finished with Ball asking her students to write in their notebooks what they thought the answer to the problem would be. The variety in students' comments reveal further the size of the debate: two students said they agreed with Sean that $6 + \hat{6}$ was six; eight students agreed with Betsy that $6 + \hat{6}$ was zero; Mei still believed that the answer was nine; Lucy said that she did not know; and Nathan noted that the answer was six because he "just thought so." Few students were absent or did not write in their notebooks. The following day, October 13, Ball introduced the class to a new model, involving money, so as to give the students more tools to reason about integer addition.

Discussion of the Episode

The proving task with which Ball engaged her third graders in the episode was about finding and *justifying* the answer to a *single* calculation problem: $6 + \hat{6} = ?$ (line 1c). In higher grades, solving this calculation problem could reduce to straightforward application of a known procedure, thus disqualifying the problem from being considered a proving task. Yet this was not the case in Ball's class: The problem engaged the students in a cognitively demanding mathematical activity that was also characterized by a genuine search for the correct answer and for a justification (proof) of the reasonableness of that answer.

In fact one could say that the calculation problem in this episode did not just satisfy but exceeded the expectations of a typical proving task, giving rise to what Balacheff called "a context promoting awareness of the need for proof" (Balacheff, 1988b, p. 285). According to Balacheff, this is a context that "holds some risk linked to *uncertainty*, and therefore something to *gain* by entering a proving process" (Balacheff, 1988b, p. 285; emphasis added). The elements of *uncertainty* and *gain* featured prominently in the episode. Uncertainty emerged from the many different answers to the problem that were proposed and supported passionately by different students. Which of all those answers was the correct one? Six, as proposed by Sean (line 3)? Zero, as proposed by Betsy (line 4)? Or nine, as declared at the end of the episode by Mei (line 20)? The community's expected gain from entering a proving process was twofold: first, to establish the correct answer to the problem thus resolving on mathematical grounds students' disagreement over the correctness of their proposed answers; second, to help students understand why only one of the answers was correct. These expected gains relate, respectively, to the two primary functions of proof discussed in Chapter 2: justification/refutation and explanation.

Although the episode presented a context for promoting an awareness of the need for proof, it did not present an actual proof and it finished with a wide range of opinion among students about what the correct answer to the problem was. The main reason for students' difficulties in solving the problem was the lack of the concept of addition of a negative integer (as the second addend in an addition sentence) from the community's *set of accepted statements.* A couple of days before the episode (on October 10) Lucy had made a promising attempt to articulate how the class might deal with addition of a negative integer in the particular case of $\hat{8}+\hat{8}=?$, but the lack of the concept of addition of a negative integer from the community's shared knowledge was nevertheless evident in the episode. Specifically, the lack was evident in the following: the contrasting views expressed by different students about whether the second addend in $6+\hat{6}=?$ indicated "plus" or "minus" (lines 10, 11, 12, and 19); students' different proposals about how to deal with this addend, with Sean on the one hand insisting on ignoring it (lines 13, 15, and 17) and Betsy on the other hand suggesting its interpretation as "minus six" (line 9); and the overall heated tone of the discussion, with several exchanges between pairs of students questioning each other's views and requiring from each other sensible explanations for views that they themselves considered to be unreasonable (e.g., lines 9–10, 13–14, 15–16, and 17–18). Despite the absence of a proof, the students' discussion included some interesting arguments, such as Sean's *non-genuine mathematical argument* about why one should ignore the term $+\hat{6}$ (lines 13 and 15) or Betsy's tacit argument by *reductio ad absurdum* about why it would not make sense to interpret this term as "plus six" (line 18).

Of course there was no definite logical resolution of the issue debated by the students about how to conceptualize addition of a negative integer, for the current rules of working with negative integers are conventions with a long and turbulent history and do not follow cogently from whole number arithmetic (e.g., Kitcher, 1984; Kline, 1972). Also, one could not expect that the students in the episode would be able to reach, or come close to reaching, such a resolution without a decisive intervention from the teacher, for students' learning processes of integer operations evolve slowly and include traces of the obstacles encountered during the historical development of these operations (e.g., Hefendehl-Hebeker, 1991; Sfard, 1991).

Thus, there was a need for teacher intervention so that the class could develop its set of accepted statements to include a more expanded concept of integer addition. That could allow, then, the formulation and acceptance of a proof for zero as the correct answer to $6+\hat{6}=?$. Ball began to organize such an intervention by introducing the class to a new model involving money. The new model was presumably more suitable than the building model to represent integer addition where the second addend in a two-addend sum was negative. According to Linchevski and Williams (1999, pp. 134–135):

> models [for working with positive and negative numbers] must describe a reality that is meaningful to the student, in which the extended world of negative numbers already exists and the students' activities allow them to discover it. This world must include the practical need for two sorts of numbers, and the relevant laws must be deducible without mental acrobatics.

It is not my goal to discuss the pedagogical potential of the money model and the extent to which it satisfies, or can satisfy, the requirements described by Linchevski and Williams. I simply note that this model has the potential to allow representation of a negative integer as a debit made to an account, a positive integer as a credit made to an account, and addition of two or more integers as the action of factoring in the respective debits or credits to find the account balance. If this way of conceptualizing integers and integer addition belonged to a classroom community's set of accepted statements, then it could not be difficult for the community to prove that $6+\hat{6}$ must be zero. Such a proof might state, for example, that an account with a credit of 6 dollars (or another unit of money) and a debit of 6 dollars made to it would have a zero balance, as the equal amounts of credit and debit would cancel each other out.

To conclude, the proving activity in which Ball engaged her class in this episode took place in the context of a task involving a single calculation problem that required students to find and justify the sum of two integers with the second one being negative. The episode presented a situation where there was an evident and pressing need for a proof, with students debating over the correctness of different answers to the problem and trying to make sense of what is involved in adding a negative integer. At the same time, however, the community's set of accepted statements did not support the development of a proof due to its lack of an essential element, namely, the concept of addition of a negative integer. Thus the proving task afforded students with an opportunity not so much to construct arguments from accepted statements, as is typically the case with proving tasks, but rather to explore and debate the foundational knowledge on which relevant arguments could be constructed. This exploration and debate also revealed the constraints of the existing knowledge and indicated the need for its expansion. The teacher played a key role in creating a safe environment for students' exploration and debate, having cultivated in previous lessons the dialogic discursive practices evident in this episode; a similar environment and set of discursive practices were illustrated in the episode from Zack's class in Chapter 1. The teacher also had an important role to play in supporting the development of the community's set of accepted statements so as both to enhance students' concept of integer addition and to create a stronger foundation for proof construction in the context of similar problems in the future. The teacher began to respond to this role by introducing the class to a new model that was better suited to represent integer addition. Overall, students' engagement with proving in the episode had become the trigger for mathematical sense-making, in the form of concept development, in the domain of integer arithmetic.

General Discussion

Episodes C and D presented two contrasting classroom situations of what it might mean or look like when elementary teachers engage their students with proving tasks that involve a single case, specifically, a single calculation. In this final section I discuss some general issues: the nexus between calculation work and proving; the relationship between proving tasks and proving activity, in the particular context of proving tasks involving a single calculation; and the role of the teacher while implementing this kind of task in the classroom.

The Nexus between Calculation Work and Proving

The proving tasks used by the teachers in these two episodes were placed in the context of single calculations in the domain of arithmetic. This is a point worth reflecting upon, for the notion of proof tends to be dissociated from work in arithmetic (including calculation work) that predominates in elementary school mathematics, with proof making its (sudden) appearance in students' mathematical work in other mathematical domains typically taught in secondary school. According to a group of researchers from four different countries (the United States, England, Germany, and Israel), mathematics instruction at elementary school tends to focus "on arithmetic concepts, calculations, and algorithms, and, then, as [students] enter secondary school, [students] are suddenly required to understand and write proofs, mostly in geometry" (Ball et al., 2002, pp. 907–908).

Despite common practice, however, calculation work, or work in arithmetic more broadly, need not be separated from proving. In fact one can make the case that proving deserves a place, and can play an important role, in this work. Two aspects of this role were exemplified in the two episodes described in this chapter: Proving can afford students with an opportunity to deepen their existing knowledge of calculation algorithms (Episode C) and can become the trigger for concept development in the broader domain of arithmetic (Episode D). A third aspect of this role derives from the literature on early algebra: A more substantial place of proving in calculation work can help draw students' attention to, and foster students' understanding of, the rules and structural relations that underpin work in arithmetic and that are fundamental to algebra (e.g., Carpenter, Levi, Berman, & Pligge, 2005; Carraher et al., 2006; Russell, Schifter, & Bastable, 2011b), thus also addressing recent calls for "early algebraization" (Blanton & Kaput, 2011; Cai & Knuth, 2011). Take, for example, a proving task that asks young children, who are not yet able to carry out the calculation on the left side of the following equality, to find and justify what number they would need to put in the box to make it a true number sentence:

$$91 + 14 = 92 + \square$$

This task can encourage students to reason about the relations between the two sides of the equality, in a way that is also helpful for future algebraic work when students will have to deal with expressions that are not amenable to direct calculation (e.g., Carpenter et al., 2003).

To conclude, given how large a part calculation work occupies in the elementary mathematics curriculum of different countries, and also the powerful aspect of mathematical sense-making that proving can bring to this work, one may talk about a transformative function that an elevated place of proving in calculation work can play in elementary school mathematics. The two episodes here offered glimpses into this transformative function.

The Relationship between Proving Tasks and Proving Activity

The definition of proof that I discussed in Chapter 2 breaks a mathematical argument down into three components: the set of accepted statements, the modes of argumentation, and the modes of argument representation. The first of these components played a key role in the proving activity that was generated by the proving tasks in the two episodes. In Episode C all relevant knowledge about place value and calculation algorithms was securely included in the community's set of accepted statements, which provided students with a robust foundation to formulate a proof for the calculation problem at hand. In Episode D, however, the state of the set of accepted statements was different: It lacked essential knowledge related to integer addition, thus creating an obstacle to students' efforts to reconcile their different answers to the calculation problem at hand and to prove the correct answer. Accordingly, in Episode D the proving activity was more about exploring the unsettled foundations of a possible argument to justify a disputable result, whereas in Episode C it was more about developing a chain of reasoning from well-established foundations to justify a certain result.

It is perhaps not surprising that the set of accepted statements played such a key role in the proving activity in the two episodes. Meaningful engagement with calculations in the domain of arithmetic is heavily dependent on knowledge of relevant concepts (e.g., place value, the meaning of operations, the meaning of the equals sign) and calculation procedures (e.g., the steps of the standard written methods for the different operations). If this foundational knowledge is not readily established or shared among the members of a classroom community as they work on a proving task that involves a calculation, the proving activity is unlikely to lead to the development of a proof. But this is not a problem per se, for, as we saw in Episode D, the limitations in the set of accepted statements during a proving activity can help motivate knowledge development in the broader domain of arithmetic. The "learning residue" (Davis, 1992) at the end of a proving task that falls into this category may not directly relate to proving, but the expanded arithmetical knowledge that is expected to derive from it may enable the successful completion of similar proving tasks in the future. Students' work in Episode C is a case in point.

The set of accepted statements on which students' proving activity in the two episodes built, or could have built, included among other elements the steps of a calculation algorithm (Episode C) and the rules of a model (Episode D). These steps or rules played, or could have played, the role of "local axioms" (Freudenthal, 1973) in students' work, with students drawing freely on them in developing their arguments (for a definition of "local axioms," see Table 2.1). The rules of a model in particular would not normally form the axiomatic foundation of a mathematician's proving activity, but it is nevertheless pedagogically defensible to allow these rules to serve as local axioms in students' proving activity. This position finds support from Freudenthal's (1973) work and that of other scholars (e.g.,

Poincaré, 1914/2009; Wu, 2002), with Poincaré (1914), for example, arguing that it is important for students "to learn to reason exactly with the axioms once admitted" (p. 136) rather than for them to admit the same principles accepted by mathematicians. Indeed, the definition of proof presented in Chapter 2 allows local axioms to form part of a classroom community's set of accepted statements (Table 2.1).

Finally, although my focus here has been on a special category of proving tasks involving a single case—those placed in the context of a single calculation—the issues I have discussed may also apply for other categories of proving tasks involving a single case. I just mention briefly another example of a proving task deriving from a different episode of Ball's class that was discussed elsewhere (Ball & Bass, 2000b). In that episode the students were debating over a claim by Sean that the number six is "even and odd." The justification or refutation of Sean's claim, which played the role of a proving task involving a single case, generated proving activity where mathematical definitions, an element of the set of accepted statements, featured prominently in classroom work. Several students tried to refute the claim by appeal to the classroom community's definitions of even and odd numbers: The number six fulfilled the definition of an even number but not the definition of an odd number. However, as it turned out, Sean had in mind a special family of even numbers, those that have an odd number of groups of two ($\equiv 2 \bmod 4$), but he lacked the mathematical language to appropriately describe them. We thus see, again, the important role played by the set of accepted statements in students' proving activity. In particular we see how the expansion or refinement of this set can support the resolution of a classroom debate associated with proving or motivate more broadly the development of a classroom community's mathematical knowledge.[3]

The Role of the Teacher

These episodes illuminate different aspects of the role of a teacher who implements a proving task involving a single case, specifically, a single calculation. This relates to the teacher considering the community's set of accepted statements with regard to both the proving task at hand and the learning goals that the teacher might want to promote at the given time.

Episode D presented a situation where the set of accepted statements did not support the development of a proof for the given proving task. A possible course of action the teacher could follow in this situation would be to lead the work of the class to a closure and inform the students that it would not be possible to resolve their debate over the correct answer to the task, because the class did not have yet the necessary knowledge to show that an answer was correct (Stylianides, 2007b). This course of action would be particularly relevant in situations where the expansion or refinement of a community's set of accepted statements was not practically possible or within the teacher's short-term plan. However, the teacher in the episode was not in this situation, and she followed another course of action: She initiated

[3] This episode relates also to the discussion in Chapter 4 about the nexus between assumptions and definitions. In particular, the debate between Sean and other students in the class over his claim seemed to be underpinned by different assumptions about the choice of definitions and about whether there was any flexibility in using already defined terms to describe a new family of numbers.

an intervention introducing the class to a new model, aiming to equip the students with the necessary knowledge about integer addition that would allow them eventually to find and justify the correct answer to the given task or other similar tasks.

Ball did not rush into introducing the new model. Rather, she first provided her students with an opportunity to engage in discussion and debate about what might be involved in adding a negative integer, thus triggering an "intellectual need" (Harel, 1998) among them to learn more about integer addition. By so doing Ball also served another important learning goal: she engaged the class in *argumentation*, a term which is generally used to describe the discourse or rhetorical means (not necessarily mathematical) used by an individual or a group to convince others that a statement is true or false (e.g., Boero et al., 1996; Krummheuer, 1995; Mariotti, 2006). For example, the rhetorical means used by Sean to convince the rest of the class that the term $+\hat{6}$ in the sentence $6+\hat{6}=?$ "would just disappear" (line 15) are not mathematical, but they nevertheless fit in with the meaning of argumentation.

Episode C presented a rather unproblematic situation for the teacher: The classroom community's set of accepted statements was fully conducive to the development of a proof for the given proving task and the students were drawing competently on that foundation in developing their arguments. As the expansion or refinement of the community's set of accepted statements was not an issue, the teacher used the proving task as an opportunity to solidify knowledge that had already been established in the class. Thus her role in the episode was less interventionist and more facilitative. For the most part, she asked probing questions, which, similar to their function discussed in the literature (e.g., Sahin & Kulm, 2008), encouraged students to think more deeply about the arithmetical concepts that underpinned their arguments.

6

Proving Tasks Involving Multiple but Finitely Many Cases

In this chapter I examine proving tasks involving multiple but finitely many cases, with an emphasis on the proving activity that this kind of task can help generate in the classroom and the role of the teacher while implementing the tasks. Specifically, I focus on proving tasks that engage students in finding all possibilities involved in a situation (multiple but finitely many) and justifying that all possibilities have indeed been found. While this specific category of tasks does not cover the full range of proving tasks involving multiple but finitely many cases, it does nevertheless relate to many different families of problems that are quite common in elementary school mathematics, such as combination, permutation, and Cartesian product problems.

As in Chapters 4 and 5, I situate my discussion in the context of two classroom episodes: the first (Episode E) comes from Ball's third-grade class in the United States (8–9-year-olds); the second (Episode F) comes from Howard's Year 4 class in England (again 8–9-year-olds). These episodes have been selected for their illustrative power and for the complementary issues they raise about what it might mean or look like when elementary teachers use in their classes this particular kind of proving task. I describe and discuss each episode separately, and I conclude with a more general discussion.

Episode E[1]

Description of the Episode

This episode took place on September 12, which was the second day of the school year in Ball's third-grade class. On September 11 the class had worked on the following task:

1. *Two-Coin Problem*: "I have pennies, nickels, and dimes in my pocket. Suppose I pull out two coins. How much money might I have? [Follow-up questions:] How do you know that you have found all different amounts? Prove your answer."

[1] In describing and discussing this episode I use with permission parts of Stylianides and Ball (2008) (license number 3454130366327).

Proving in the Elementary Mathematics Classroom.

After the students had generated, as a group, all the different amounts of money one might have (i.e., six), Ball asked the follow-up questions in an attempt to challenge students to *prove* that they had actually found all possibilities. Ball wrote in her journal at the end of the lesson that day (September 11, p. 7):

2. When kids told me they were done, I *did* ask how they knew they were done. This was the question I tried to provoke near the end of class too, when we'd (as a group) listed all six combinations. But no one could say for sure that that was all. It's just that they couldn't find any other combinations. I asked them to think about it until tomorrow —was there a seventh answer? [...] The kids did not seem particularly worried about whether they'd exhausted all the possibilities or not—I think they thought they had and weren't particularly concerned with proving that. [emphasis in original]

A discussion of the work of the class on the Two-Coin Problem can be found in Stylianides (2007b, pp. 302–306). Here I focus on the lesson that took place the next day. In planning for that lesson, Ball puzzled in her journal (September 12, p. 9) over what would be a good task to use with her class:

3. Would permutations of numerals help the kids develop a sense both for a need to keep systematic track of their explorations and also how to argue that one has uncovered all possibilities? Maybe do permutations of 2, 3, 4? Then return to [the] coin problem tomorrow?

Ball ended up using the following task:

4. *Date Problem*: "Today's date is 9/12. Take the 9, the 1, and the 2. How many 3-digit numbers can you make? [Follow-up questions:] How do you know you have found them all? Prove your answer."

The lesson on September 12 began with Ball reminding the students of the work they had done the day before, setting up their work on the Date Problem:

5. *Ball*: Yesterday, when we were working on the problem with the coins, we found six solutions to our problem. Six different amounts of money that you could get. And I asked you how you knew that you had all the answers you could find, and nobody was too sure. Today we are going to do a couple of problems to help us think a little bit more about telling when you have all the answers. And the problems are a little bit different than the ones we worked on yesterday. [...] Just to kind of give you the idea of a problem that you're going to work on for a few minutes, in a few minutes, I want to sort of warm up with a smaller problem. Here's the problem I want you to think about: Suppose you had the number 5 and you had the number 9 and you wanted to make a two-digit number. What number could you make with a 5 and a 9?

Ball began the class with a problem about permutations of two digits, 5 and 9, as a warm-up for the Date Problem. Lucy and Lisa provided almost right away the two possible permutations: 59 and 95. Ball emphasized that, in working on problems like this one, the class would focus on *arrangements* of digits, which she described as "sliding the numbers together." The students then worked in their small groups for about 20 minutes on the Date Problem (which was presented to them as in line 4 but without the follow-up questions yet).

In the discussion that followed, the class generated, as a group, six permutations. Ball pushed the students to show whether they had actually found *all* three-digit numbers they could make:

6. *Ball*: Lucy says there are only six of them. *How do you know that we have them all?* How can we be sure that we have them all? [...] Mei, what do you think?
7. *Mei*: Well, I think we have them all, because I kept on doing numbers and trying them until they were the same ones.
8. *Ball*: Jeannie, what do you think about Mei's reason? She said she kept doing them, and she kept getting the same ones and that's why she thinks this is it.
9. *Jeannie*: I don't think, I think she is right, that um, there isn't any more.
10. *Ball*: Is there any way to *prove* that we've gotten them all? If somebody said, "How do you know you've got them all?" Is there a way you could *prove* it so the person would have to believe you? Lucy?
11. *Lucy*: Um, people would know that we did them all because people started saying them over.
12. *Ball*: Okay, so that's kind of like what Mei said, right? You just keep doing them; when we start saying the same thing, we must have them all. Betsy?
13. *Betsy*: You can prove by, you take two 9s, so you take two 9s at the beginning.
14. *Ball*: Come up and show us, I'm not sure what... [She turned to the class.] Watch now and see what you think of what Betsy's...
15. *Betsy*: Two 9s, two 9s, okay then you put the 2 first here and the 1 first there and then you put the 1 and the 2 and you get two different ones. [As she talked, she wrote on the board, digit by digit, the following permutations: 921 and 912.] And then you take the 2 first, two spots, and then you take the 9 and put it somewhere and you take the 1 and put it somewhere. And then you put the 1 here and the 9 there and you get both, you can't get anymore. [As she talked, she wrote on the board, digit by digit, the following permutations: 291 and 219.] Then you take the 1 first, and then you get the 9 here and the 2 there. And then you get the 2 here and the 1 there and that's all you can get. [As she talked, she wrote on the board, digit by digit, the following permutations: 192 and 121.]
16. *Students*: The 9! [They noticed that in the last permutation, 121, the last digit had to be 9.]

17a. *Betsy*: Put the 9 there [Betsy corrected her mistake] and that's all you can get, 'cause you're using in different spots each time.

These were the six permutations that were written by Betsy on the board:

921	912
291	219
192	129

Betsy then moved on to review again part of her argument by referring to the first two permutations: 921 and 912.

17b. *Betsy* (cont.): 'Cause it's first here and last there [she pointed at the place of 2 in 921 and 912], first here [and] last there [she pointed at the place of 1 in the same two permutations], and you can't get any more with that one number. Using these two [the digits 1 and 2], you can only put them at the end [she referred to the last two places of the two permutations]. Two times a certain way and with these two when they're at the beginning [she referred to the place of 9 in the two permutations], you can only put these two [the digits 1 and 2] at the end in different ways.

Ball invited comments from the rest of the class. Lisa asked Betsy:

18. *Lisa*: How did you figure that out?
19. *Betsy*: I just thought of it.
20. *Ball*: What do you think about it, Lisa?
21. *Lisa*: I just thought that was a really neat idea.
22. *Ball*: Does it make sense?
23. *Students*: Uh huh.

The discussion continued with Ball asking some questions to help more students understand Betsy's argument. The students who talked seemed to have grasped the idea of the argument, and then Ball introduced a new task:

24. *Ball*: What would happen if we took three people instead of three numbers and had them line up? Suppose Mrs. Rundquist said um, I don't know, Haroun and Chris and Sean could go to the library if they lined up in a straight line and then they could go to the library again the next day if they could line up in a different line, in a different order. And they could keep going to the library every day until their line, they made a line that they had made before. How many lines do you think those three kids could make?

The students began to work on this task and then on an extension of it that talked about *four* students lining up in different arrangements. The lesson ended shortly thereafter.

Discussion of the Episode

The proving tasks with which Ball engaged the class in this episode and also during the first week of the school year are summarized in Table 6.1. All of these tasks were about finding and *justifying* that all possibilities involved in a situation have been found, with the number of possibilities being *finite* and relatively small. The issue of proof, which is reflected in the follow-up questions for each task, was not raised when the tasks were first presented to the students. Rather, the teacher raised this issue after the students had an opportunity to generate different possibilities and (many of them) thought they had found them all.

There is some difference between the situations described in the tasks in Table 6.1. The Coin Problems describe a situation that involves combining the values of different coins selected from a pool of three different coins (penny, nickel, dime), allowing duplicates (e.g., two pennies is a viable combination for the Two-Coin Problem) but disregarding different orderings (e.g., a penny and a nickel does not count as a different combination from that of a nickel and a penny). This situation relates to a family of problems often called "combination problems with repetition." The Date and Lining Up Problems describe a situation that involves arranging into order a number of distinct objects: digits and students, respectively. This situation relates to a family of problems often called "permutation problems without repetition." These families of problems lend themselves to different efficient methods for enumerating all possibilities.[2] However, the specific problems used by Ball,

Table 6.1 Proving tasks used during the first week of the school year in Ball's class.

Dates	Proving tasks[a]
September 11	*Two-Coin Problem*: "I have pennies, nickels, and dimes in my pocket. Suppose I pull out two coins. How much money might I have? [Follow-up questions:] How do you know that you have found all different amounts? Prove your answer."
September 12–13	*Date Problem*: "Today's date is 9/12. Take the 9, the 1, and the 2. How many 3-digit numbers can you make? [Follow-up questions:] How do you know you have found them all? Prove your answer." *Lining Up Problem*: "Mrs. Rundquist said that Haroun, Chris, and Sean could go to the library if they lined up in a straight line. They could go to the library again the next day if they could line up in a *different* order. And they could keep going to the library every day until they made a line they had made before. [Follow-up questions:] How many different lines could those kids make? How do you know you have found them all? Prove your answer." *Extended Lining Up Problem*: The same as above but with *four* students lining up.
September 18	*Three-Coin Problem*: "I have pennies, nickels, and dimes in my pocket. Suppose I pull out *three* coins. How much money might I have? [Follow-up questions:] How do you know that you have found all different amounts? Prove your answer."

[a]The teacher raised the issue of proof, which is reflected in the follow-up questions for each task, after the students had an opportunity to generate different possibilities and (many of them) thought they had found them all.

[2] The number of combinations in the Coin Problems is given by the formula $\binom{n+k-1}{k}$, where n represents the number of coins to choose from (which would be equal to 3 in both problems) and k represents the number of coins selected (which would be equal to 2 in the Two-Coin Problem and to 3 in the Three-Coin Problem). The number of permutations in the Date and Lining Up Problems would be given by the formula $n!$, where n represents the number of distinct objects being arranged into order (which would be equal to 3 in the Date and Lining Up Problems and to 4 in the Extended Lining Up Problem).

which illustrate problems that are rather common in elementary school mathematics (save for the expectation for a proof), do not require knowledge or use of efficient enumeration methods for their solution. The relatively small number of possibilities involved in these problems makes them accessible to elementary students who can solve the problems using the mode of argumentation associated with the systematic enumeration of all possibilities.

Indeed, this was the mode of argumentation targeted by Ball when she engaged her students with these particular proving tasks. As she wrote in her journal, she wanted to help her students "develop a sense both for a need to keep systematic track of their explorations and also how to argue that one has uncovered all possibilities" (line 3). She also communicated this goal to the students at the beginning of the lesson on September 12, when she said, "Today we are going to do a couple of problems to help us think a little bit more about telling when you have all the answers" (line 5).

The common mode of argumentation called for by the solution of the tasks helps explain, at least partly, the similarities in the proving activities (including student difficulties) that were observed during the implementation of the tasks that Ball used during the first two days of the school year. Consider, for example, the beginning of the whole class discussion of the Date Problem on September 12, which was very similar to the discussion of the Two-Coin Problem the day before (cf. line 2). The students generated, as a group, all six possibilities and they were convinced that they had found all of them. The students' conviction had no robust mathematical grounds and was questioned by Ball (line 6) who challenged them to prove their claim that they had indeed found all the possibilities (lines 6 and 10).

Mei provided an argument involving *un*systematic enumeration of cases: She said she knew she had found all the possibilities because she kept trying different arrangements of the three digits until they started repeating (line 7). This was an *empirical argument* that used an invalid mode of argumentation and offered inclusive evidence for the claim that these were indeed all the possibilities. Ball, unsatisfied with this argument, continued to play the role of the "skeptic" (Mason, 1982) and pushed the students to come up with a way to *prove* their claim (line 10). Lucy then offered another empirical argument (line 11). Ball pointed out the similarity between the arguments of Lucy and Mei, thus indicating to the students that they had not yet developed a proof (line 12). At this point, Betsy volunteered to offer a proof by crafting a rule for systematization (line 13).

According to the definition of proof described in Chapter 2, Betsy's argument (lines 15 and 17a, b) qualified as a proof for the given task because:

(1) It used true statements that were readily accepted by the class, such as the statement that, with the digit 9 as the first object in the permutation, there were only two ways to arrange the remaining two digits (in one of the ways the digit 1 is the second object and the digit 2 is the third, and in the other way the place of the two digits is reversed).

(2) It employed the valid mode of argumentation associated with enumerating systematically all possibilities involved in a situation (permutations of three distinct digits) that was within the reach of the classroom community.

(3) It was represented appropriately using a combination of verbal and written language that was understandable to the students in the class.

It is encouraging to observe that Betsy's argument was appreciated and accepted by the class (lines 18–23), with Lisa characterizing it "a really neat idea" (line 21).

An issue that may arise during the implementation of proving tasks like the Date Problem (particularly tasks that involve permutations of a larger number of objects) is how to *refute* false statements. Consider, for example, how an elementary student might refute the statement "there are exactly two permutations of the digits 1, 2, and 9." A possible argument could be based on strategic enumeration of cases, that is, the enumeration of more than two permutations, say three. Another possible argument could build on the following observation: Given that one gets two permutations with *two* distinct digits, how can it be possible that there are still two permutations with *three* distinct digits? An argument by *reductio ad absurdum* that would build on this observation would assume a monotonic relationship between natural numbers n and the number of permutations of n distinct elements, which, although intuitively true, might require justification. Note also that the previous two arguments show that there are more than two permutations, but they do not explain why there are exactly six of them.

Finally, this episode suggests that teachers who design or implement the same kind of tasks as Ball did during this episode should be prepared for students to have difficulties related to understanding or using the valid mode of argumentation associated with the systematic enumeration of all possible cases involved in a situation. As illustrated by this episode, students may tend to use empirical arguments and not readily see an "intellectual need" (Harel, 1998) for alternative arguments that meet the standard of proof (from the point of view defined in this book), thus creating challenges for teachers as they manage their students' proving activity.

Episode F

Description of the Episode

This episode took place during November 5 and 6 in Howard's Year 4 class. Howard started the lesson on November 5 by saying that the class would work on "How many ways . . . " problems. Specifically, she explained that the class would work on finding how many different outfits Professor McGonagall[3] could have using different hats, dresses, etc. The first task was as follows:

1. *Outfits Problem*: "How many ways are there to dress Professor McGonagall? She has 3 different dresses and 2 different hats [see Figure 6.1]. (Prove your answer.)"

Although the text in parentheses was not an explicit part of the statement of the task, it was understood in the class, and became clear during the task's implementation, that the teacher did not just expect students to find the number of different ways there were, but also to prove their answers.[4]

[3] This is a character in the Harry Potter book series (e.g., Rowling, 1997).

[4] The expectation for a proof was explicitly stated in the two extensions of the Outfits Problem (see lines 22 and 25).

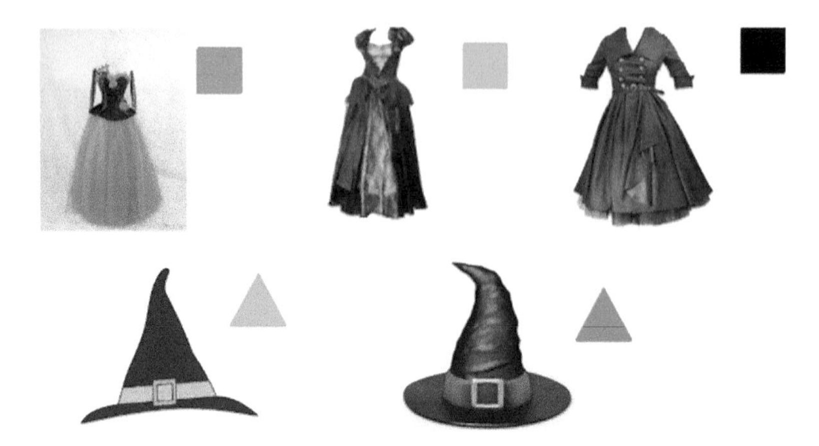

Figure 6.1 The three different dresses and two different hats of Professor McGonagall in the Outfits Problem.

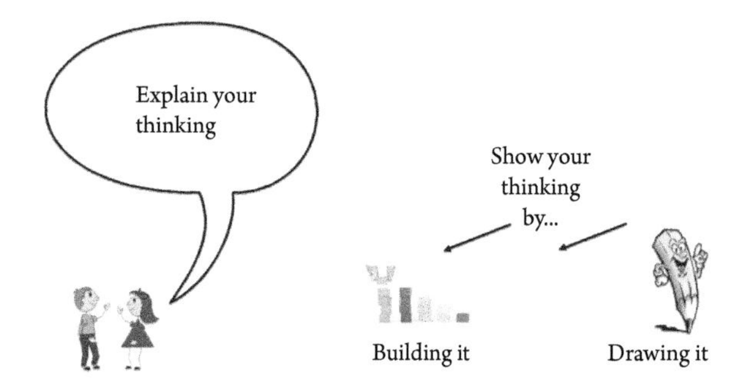

Figure 6.2 The slide used by the teacher to supplement her instructions for the Outfits Problem.

Howard allowed a couple of minutes for the students to discuss their thoughts about the problem with a person sitting next to them. Then she invited some feedback from that discussion, clarifying that she wanted students to say "something interesting, not the actual answer." Some students observed that one could combine different hats with different dresses, mentioning also specific examples of combinations.

Then Howard gave to the class the following instructions, using also the slide that is presented in Figure 6.2.

2. *Howard*: Explain your thinking and show me how many different outfits there are by drawing it on the piece of paper or by building it using pattern blocks, multi-link cubes... [These resources were made available to students who could use them if they wanted to.]

Some students asked the teacher a few clarifying questions, such as whether they needed to draw the actual dresses. Howard replied that they did not need to make complicated drawings, because, then, they "wouldn't be doing maths."

During the small-group work that followed, about half of the students in the class were making drawings and the rest were using resources from those available. For example, two boys used multi-link cubes: three cubes of the same color on the top of each other to represent a dress and then another cube on the very top to represent a hat. The students who chose to do drawings were making slower progress, partly because they tended to produce detailed and rather realistic illustrations of the outfits (see, e.g., Samantha's initial drawing of an outfit in Figure 6.3).

At some point Howard interrupted the small-group work to address a question asked by Irene:

3. *Howard*: Irene has a question. Irene, can you say the question?
4. *Irene*: Can it be a dress without a hat?
5. *Howard*: Technically it can be, but for this activity a dress always goes with a hat.

Then Howard asked the students to walk around the room to see what others had built or drawn. When the class came back together, Howard invited comments from students about what they had observed.

6. *Sonia*: Some people have only just started!
7. *Howard*: What's taking the time?

Figure 6.3 Samantha's initial drawing of an outfit for the Outfits Problem.

8. *Sonia*: The dress drawing is taking a lot of time.
9. *Howard*: What do you suggest?
10. *Sonia*: We can use a triangle and a square on the top.
11. *Howard*: What are these for?
12. *Osborn*: The triangle is for the dress and the square on the top for the body part.

After some more discussion, Howard asked everyone to use two simple shapes to represent the dress and the hat in an outfit. Then she asked the class to reflect on her reasons for doing that.

13. *Howard*: Why do I ask you to draw them like this?
14. *Orrin*: Because it will save time and we will get more maths done.

Howard gave the students three more minutes to finish their solutions. Following that, she asked the students to have another look at other people's work and see whether they could find a complete solution to the problem.

15. *Ella*: I think Cameron and Simon have all the different ways.
16. *Howard*: How many do they have?
17. *Ella*: Six.

Howard showed Cameron and Simon's solution to the class, which was similar to Irene's solution that is presented in Figure 6.4. Other students nominated Jim and Andy's solution using multi-link blocks, which is shown in Figure 6.5(a). Howard pointed out the simplicity of the representation in this and the previous solution. She also proposed that Jim and Andy could have used only one cube, instead of three, to represent a dress. She drew what she meant on the board, as shown in Figure 6.5(b).

18. *Howard*: Does it look like a dress [she referred to her drawing]?
19. *Students*: No!
20. *Howard*: Does it matter?
21. *Students*: No!

Then Howard presented the class with a new task, which was an extension of the previous one.

22. *Extended Outfits Problem*: "Now Professor McGonagall has 3 different dresses and 3 different hats. Think how many different outfits there are now and prove it to me!"

All the dresses and hats available for the new task were shown on a slide, which was as in Figure 6.1 with the addition of a purple hat. The students worked on the new task in their small groups.

Figure 6.4 Irene's solution to the Outfits Problem.

Osborn, who was sitting close to me, wrote almost immediately on his paper: "We think there is 9 [outfits] with 3 hats." I asked him how he figured out the answer so quickly. He said:

23. *Osborn*: You need to combine all hats with all dresses. The new hat is going to give three new outfits, which you need to add to the six you had before.

Osborn's reasoning was consistent with Eve's solution to the Extended Outfits Problem that is presented in Figure 6.6.

Orrin, who was sitting next to Osborn, joined my conversation with Osborn and expressed a similar thought about how the solution to the new problem could be derived from the solution to the previous one:

24. *Orrin*: There are already six and you have to put the new hat on each dress. This will give you three more.

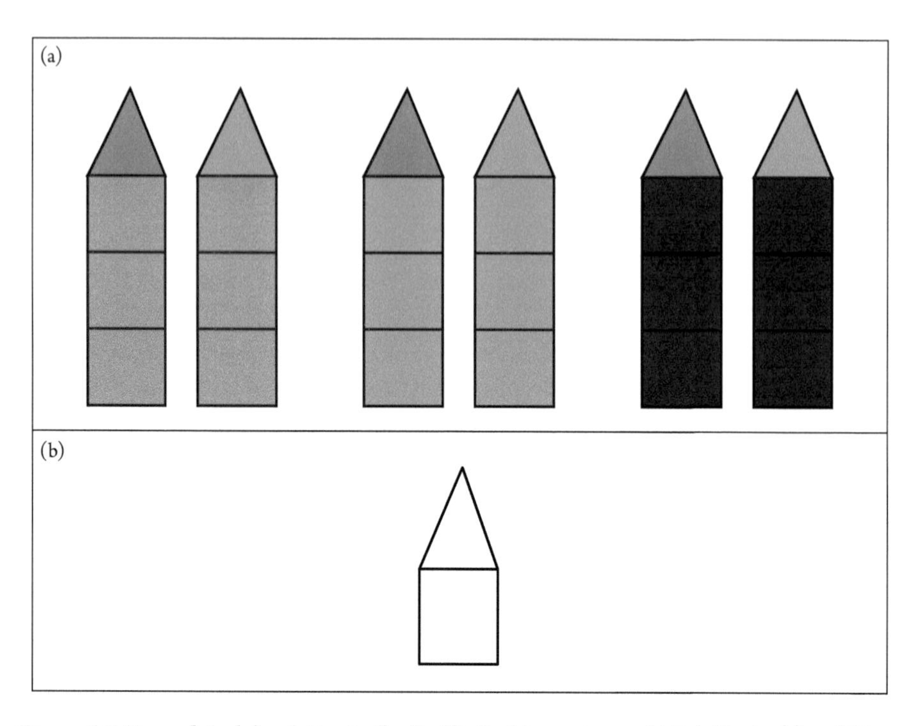

Figure 6.5 Jim and Andy's solution to the Outfits Problem using multi-link blocks (a) and Howard's simpler representation of an outfit (b).

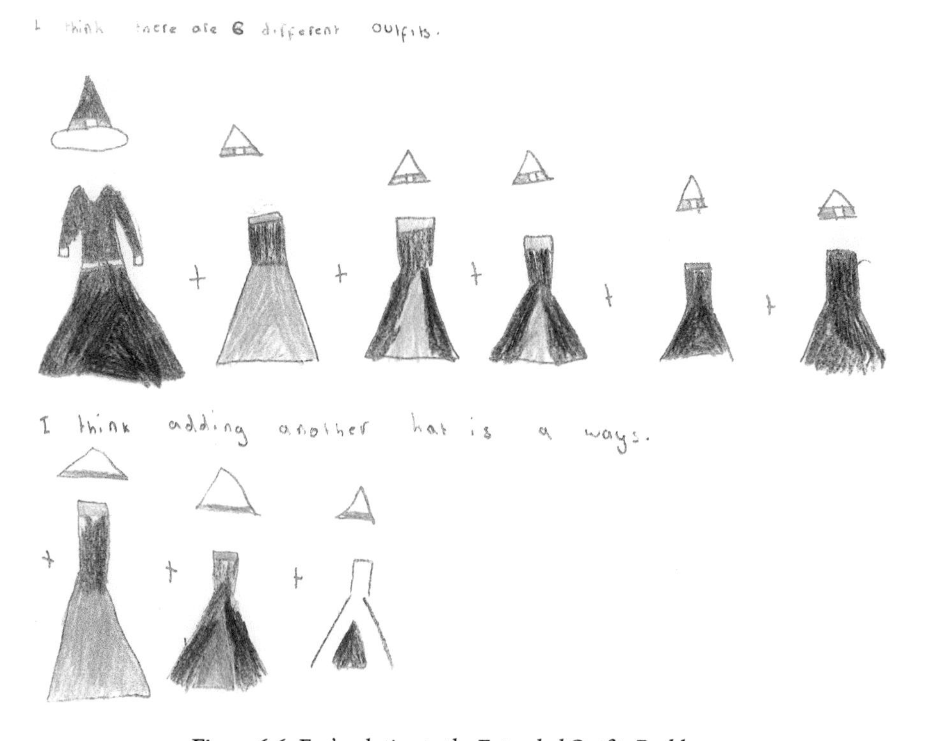

Figure 6.6 Eve's solution to the Extended Outfits Problem.

Helen's solution in Figure 6.7 illustrates another way of representing the three new outfits: Rather than presenting separately the three new outfits that have become possible with the inclusion of the purple hat (like Eve did in Figure 6.6), Helen incorporated each new outfit into the respective collection of outfits for each color dress.

In the meantime Howard came to Osborn and Orrin's desk. The two boys explained to her their solution, similar to what they had explained to me, and then they turned to use multi-link blocks. Following Howard's earlier suggestion about how to efficiently represent an outfit (Figure 6.5b), they only used two blocks for each outfit: one block for the dress and another one for the hat.

During the whole class discussion Howard asked the students to report their answer to the Extended Outfits Problem. Everyone seemed to agree that the answer was nine different outfits. Howard then presented a further extension to the task, which also included pairs of socks that could be used in an outfit.

25. *Further Extended Outfits Problem*: "Now Professor McGonagall has 3 different dresses, 3 different hats, and 3 different pairs of socks [on a slide the students could see the three pairs of socks: one green, one yellow, and one white]. Think how many different outfits there are now and prove it to me!"

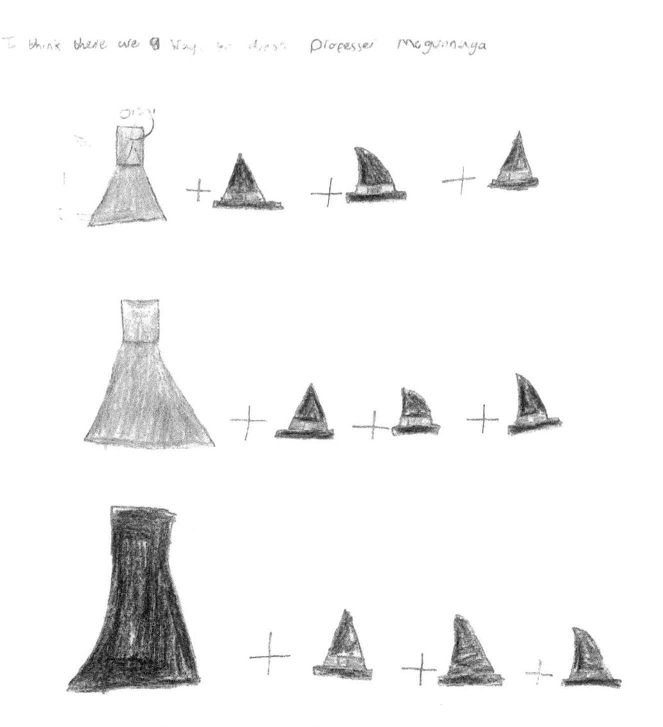

Figure 6.7 Helen's solution to the Extended Outfits Problem.

The following discussion took place:

26. *Howard*: What do you think will be the answer to this task?
27. *A student*: Twelve.
28. *Another student*: Twelve. They go up in the 3-times table.
29. *Keith*: It's going to be 18.
30. *Howard*: Okay, think more about this in your groups.

After a couple of minutes Keith and Alex discovered enthusiastically that the answer would be more than 18, and they asked the teacher to come to their table. In the meantime, Osborn was also heard to say that the answer would be more than 18, and I went to talk to him. I asked why he thought the answer would be more than 18, and after some thinking he said the answer would be $18 + 9 = 27$.

In the new whole class discussion, Howard asked how many students thought the answer would be 27, and eight students raised their hands.

31. *Howard*: Osborn, can you explain why you changed your mind, from 12, to 18, to 27?
32. *Osborn*: You have nine outfits for the hats and dresses, and now each of those nine can get a different kind of socks.

Osborn's explanation was consistent with the solutions that many students had produced on their papers for the Further Extended Outfits Problem. Figure 6.8 presents two representative student solutions, which show clearly how three different sock colors can be combined with each of the nine possible outfits from the Extended Outfits Problem to generate a total of 27 new outfits for the Further Extended Outfits Problem. The solution in Figure 6.8(a) was produced by Irene, whose work on the original Outfits Problem I presented earlier (Figure 6.4); the solution in Figure 6.8(b) was produced by Samantha who, as I had noted earlier, had begun the lesson drawing rather detailed and time-consuming illustrations of outfits (Figure 6.3).

In addition to offering a forum for the presentation of some student solutions based on drawings, Howard presented Alex and Keith's solution based on multi-link blocks. The two boys had not yet built up all 27 possibilities, but they started doing that in a systematic way. Howard presented their work to the rest of the class as an example of how one could build up all possibilities using a simple, three-cube representation for each outfit: one cube for the hat; one for the dress; and one for the socks.

Howard then challenged the class to come up with a way to show "really quickly" the 27 possibilities, asking the students to think how they could "represent colors without taking much time." Some students suggested using single line segments to represent the different parts of an outfit, or writing the name of each color instead of actually drawing it, or just writing the first letter of each color such as G for green. The teacher praised these suggestions and finished the lesson on November 5 by representing on the board different outfits using only the first letters of each color, such as GR for an outfit comprising a green dress and a red hat.

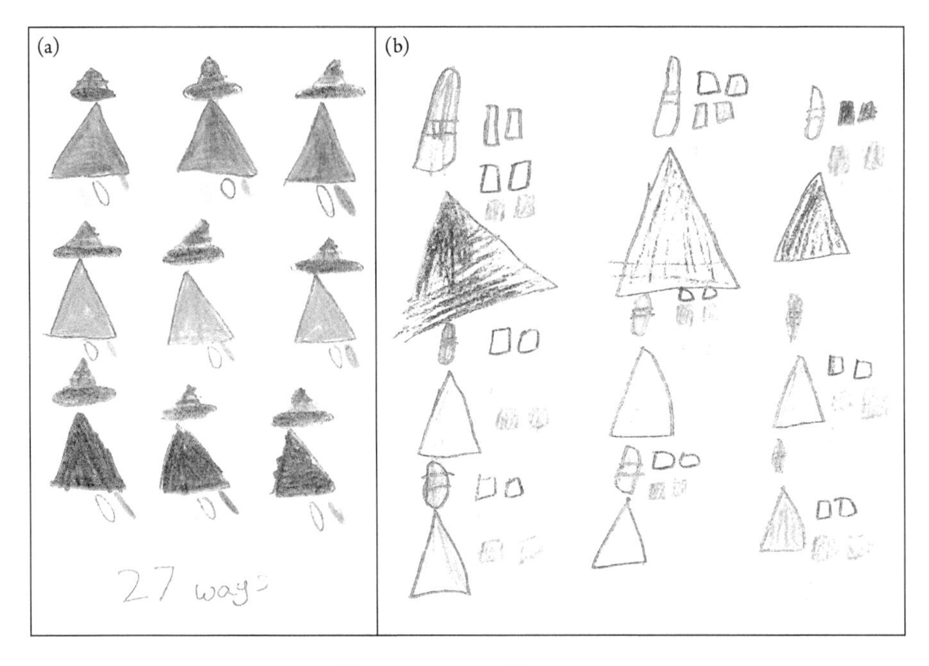

Figure 6.8 The solutions of Irene (a) and Samantha (b) to the Further Extended Outfits Problem.

The lesson on November 6 started with a review of the previous day's work and with the students sitting on the carpet holding their personal whiteboards. Howard asked the students to say what problems they had worked on. She also reminded them of her expectation that they would explain their thinking and show their work using drawings or concrete materials (the teacher again used the slide in Figure 6.2). The following discussion happened:

33. *Howard*: What was one of the problems with drawing that we noticed early on?
34. *Irene*: People made complex drawings that took a lot of time.
35. *Howard*: So we said we need simple drawings. When I went home last night I drew this. [She presented the slide in Figure 6.9].

Howard then asked the class to say what they thought the colored shapes on the slide, squares and triangles, represented. The students had no difficulty in recognizing that the squares represented dresses and the triangles hats of the respective colors.

36. *Howard*: It's like a [map] key in geography! These [shapes] shouldn't take much time to draw. But can we come up with a new code that *wouldn't* require drawing?

Figure 6.9 A slide used by the teacher to illustrate a simple way of representing the different possibilities in the Outfits Problem.

The students had a few minutes to think about this and discuss with people sitting next to them. Two boys wrote the following on their personal whiteboards:

37. *Two boys' work on their whiteboards*:

a = dress

b = hat

c = socks

ad = black dress

...

The code that these two boys started to create was obviously inefficient. In the discussion that followed, Howard chose to present to the rest of the class the work of Andy on his personal whiteboard (line 38), and she engaged Andy and other students in explaining what the different combinations of letters stood for (OD stood for orange dress, GH stood for green hat, etc.).

38. *Andy's work on his whiteboard*:

OD + GH

OD + RH

GD + GH

GD + RH

BD + GH

BD + RH

Howard presented some more examples of students' work, including the work of a girl who created a key using digits:

39. *A girl's work on her whiteboard*:

14

15

24

25

34

35

Howard then typed up on a slide the representations in lines 38 and 39, and she asked the class to compare them.

40. *Keith*: They are done in order.
41. *Howard*: What do you mean?
42. *Keith*: You can tell what's the OD [in the representation in line 38], it equals 1 [in the representation in line 39]. The GH equals 4. [...]
43. *A student*: The numbers are not random. They are a pattern.

Howard then gave to the students two options of an activity to do: They could "write the code for three dresses, three hats, and three pairs of socks" and use that code to solve the Further Extended Outfits Problem, or they could work on a new task called the Frogs Problem. The students in the class were distributed almost equally between the two options.

44. *Frogs Problem*: "Arrange three frogs [each of a different color] on their lily pads. How many different ways can you arrange three frogs? What happens if there are four frogs? (Prove your answer.)"

Models of different colored frogs were made available to students who could use them if they wanted to. While all students who worked on the Frogs Problem seemed to be clear that the three frogs had to be of different colors, a few students such as Cameron assumed that the lily pads also had to be different from each other. This assumption added a lot of complexity to the development of a code for the problem, as illustrated in Cameron's initial work on the problem that is presented in Figure 6.10.

The teacher had meant for the lily pads to be identical. It did not take long for the students who had originally made Cameron's assumption to adopt the interpretation of the problem statement intended by the teacher. In some cases this change in interpretation happened independently as students realized the complexity of their original

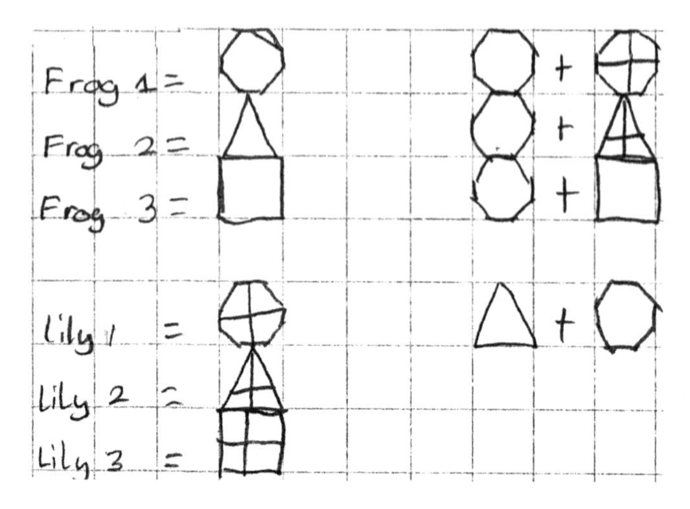

Figure 6.10 Cameron's initial work on the Frogs Problem.

interpretation, while in other cases it followed input from the teacher. By the end of the lesson Cameron himself solved the second part of the problem, which asked for arrangements of four frogs, using a code that distinguished between frogs of different colors but assumed identical lily pads (Figure 6.11). Another issue of interpretation of the problem statement came up during small-group work, with the teacher clarifying the statement for a student:

45. *A student*: Can the frogs be upside down?
46. *Howard*: This is another way to arrange them, but not for this investigation!

In Figure 6.12 I present samples of students' solutions to the first part of the Frogs Problem, which asked for arrangements of three frogs. These illustrate the variety of approaches that students followed in listing systematically all possibilities. Andy and Rachel both used codes based on digits: Andy exhausted all possibilities with 1 as the first object in a permutation before he moved on to permutations with 2 or 3 as the first object (Figure 6.12a); Rachael exhausted all possibilities with 2 as the middle object in a permutation before she moved on to permutations with 2 as the first or last object (Figure 6.12b). Irene used a letter code: She exhausted all possibilities with the letter G as the middle object in a permutation before she moved on to permutations with the remaining letters as the middle object (Figure 6.12c).

The second part of the Frogs Problem, which asked for arrangements of four frogs, naturally challenged students more than the first part. Yet there were still many students who managed to find all 24 possibilities. Some of them, including Cameron, whose solution I mentioned earlier (Figure 6.11), did not seem to apply consistently and throughout a specific rule to generate all permutations. Some others were more successful in crafting a rule for systematization. Andy's solution in Figure 6.13 showed, from the middle of the list

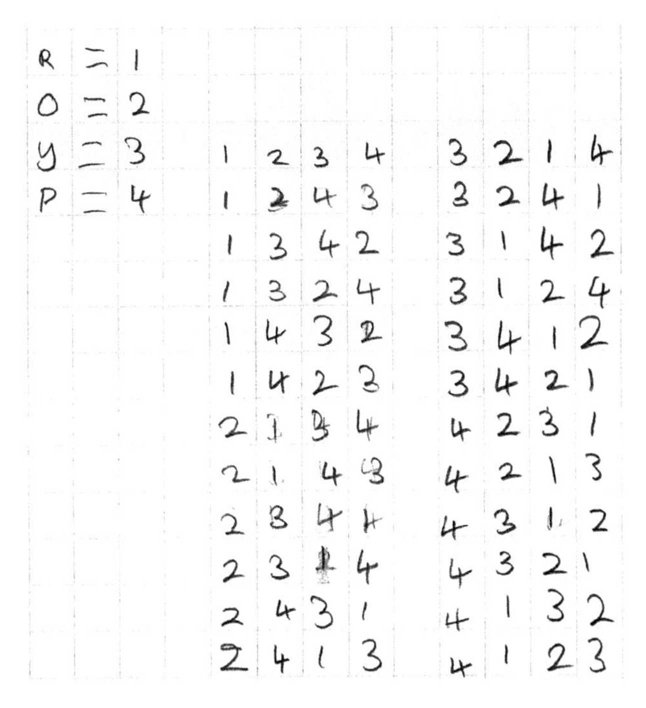

Figure 6.11 Cameron's solution to the second part of the Frogs Problem.

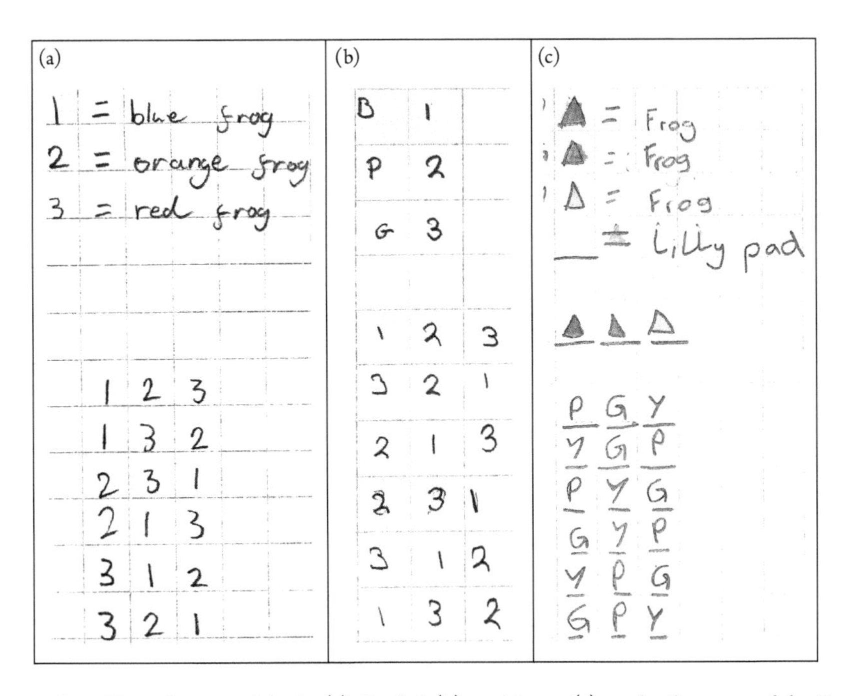

Figure 6.12 The solutions of Andy (a), Rachel (b), and Irene (c) to the first part of the Frogs Problem.

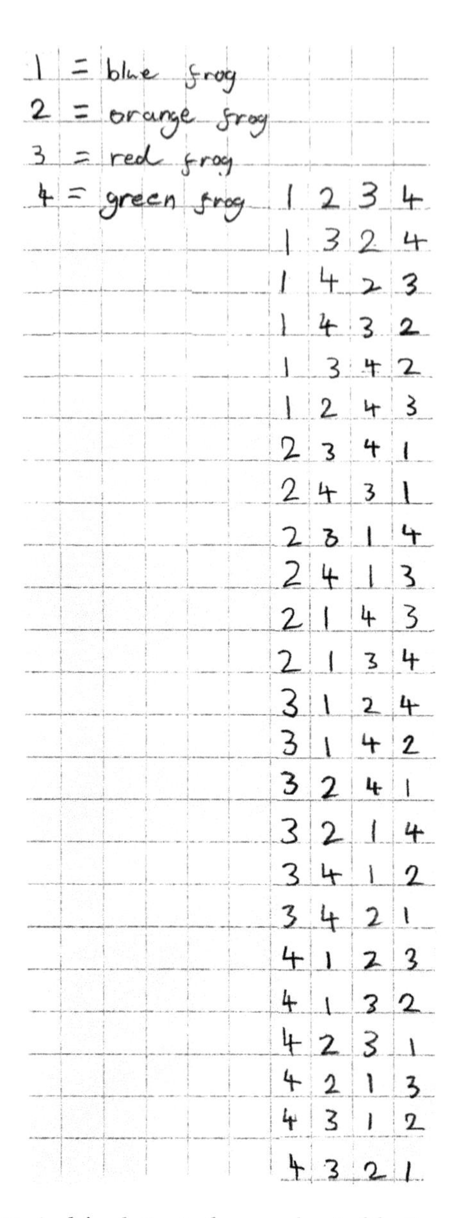

Figure 6.13 Andy's solution to the second part of the Frogs Problem.

onwards, a consistent application of a rule concerning the first two objects in the permutations. Irene's solution in Figure 6.14 showed her initial attempt to list all possibilities that led to a randomly presented and incomplete list (the left column in the figure), but this was followed by another attempt that led to a systematically presented and complete list (the middle and right columns in the figure).

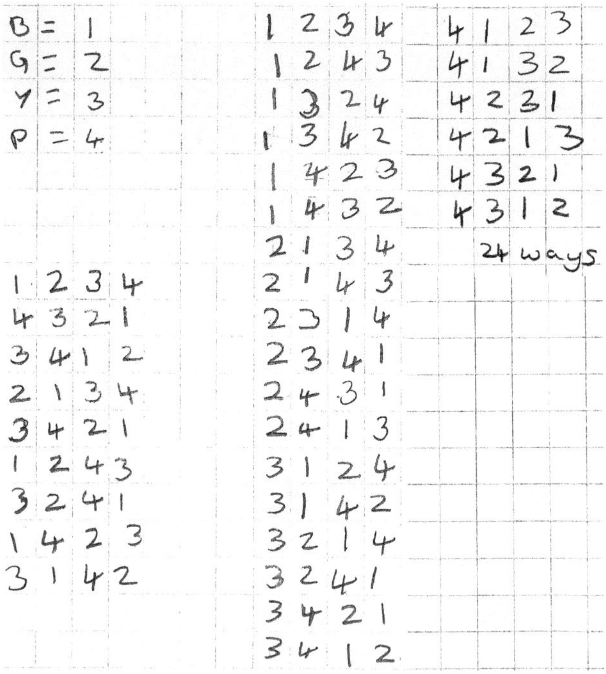

B = 1
G = 2
Y = 3
P = 4

1 2 3 4
4 3 2 1
3 4 1 2
2 1 3 4
3 4 2 1
1 2 4 3
3 2 4 1
1 4 2 3
3 1 4 2

1 2 3 4
1 2 4 3
1 3 2 4
1 3 4 2
1 4 2 3
1 4 3 2
2 1 3 4
2 1 4 3
2 3 1 4
2 3 4 1
2 4 3 1
2 4 1 3
3 1 2 4
3 1 4 2
3 2 1 4
3 2 4 1
3 4 2 1
3 4 1 2

4 1 2 3
4 1 3 2
4 2 3 1
4 2 1 3
4 3 2 1
4 3 1 2
24 ways

Figure 6.14 Irene's solution to the second part of the Frogs Problem (initial attempt in the left column; final attempt in the middle and right columns).

As I mentioned earlier, Howard had given the students the option to work either on the Frogs Problem or on a code-based solution to the Further Extended Outfits Problem. I now turn to the work of those students who chose the second option.

Some students like Cadman used letters. The code used by Cadman was not stated explicitly in his work (Figure 6.15), but most likely it included the following entries:

rh: red hat
od: orange dress
bd: black dress
gd: green dress
ws: white socks
ys: yellow socks
gh: green hat

Cadman's work was left incomplete, presumably due to lack of time. While Cadman seemed to have had a plan in mind as he was generating the different possibilities, this plan was time-consuming to execute, partly because it required him to write down two letters for each part of an outfit (i.e., six letters in total for each outfit). Also, one can reasonably

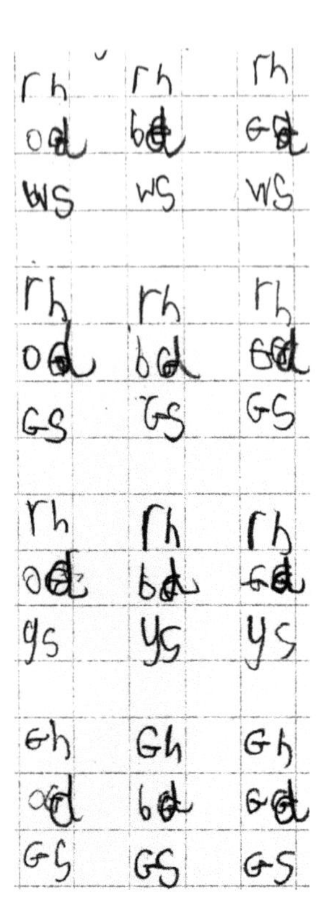

Figure 6.15 Cadman's work on the Further Extended Outfits Problem using a code based on letters.

assume that Cadman experienced, or would have experienced at some point, difficulty in keeping track which possibilities he had already covered and which others he had to write next. Antony, whose work is presented in Figure 6.16, seemed to have experienced similar challenges in working with a code based on letters. These challenges had presumably motivated him to begin to use a new code based on digits (see the top right side of his paper).

The shift from a code based on letters to a code based on digits is illustrated more clearly in the student work that is presented in Figure 6.17. This student seemed to have come up with the idea of using digits after he or she had spent some time using a code based on letters, which presumably had made the student realize the challenges involved in using a letter-based code.

Figure 6.18 presents a complete solution to the Further Extended Outfits Problem using a code based on digits (which was a re-representation of an original code based on letters). The solution reflects a systematic rule for listing all 27 possibilities.

Discussion of the Episode

In this episode Howard engaged the class in the following proving tasks: the Outfits Problem (line 1) and its two extensions, the Extended Outfits Problem (line 22) and the Further

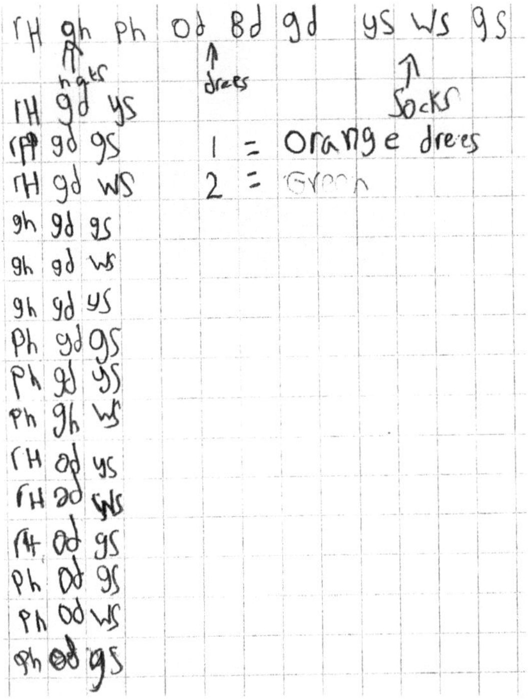

Figure 6.16 Antony's work on the Further Extended Outfits Problem using a code based on letters, and the beginning of a new code based on digits.

Extended Outfits Problem (line 25); and the Frogs Problem (line 44) that included two parts, the second of which could be considered to be an extension problem of the first. While all of these tasks are about finding and *justifying* that all possibilities involved in a situation have been found, with the number of these possibilities being *finite* and relatively small, there is some difference in the situations described in the tasks. On the one hand, the Frogs Problem belongs to the same family of problems as the Date and Lining Up Problems that I discussed in Episode E (Table 6.1); these are often called "permutation problems without repetition." On the other hand, the Outfits Problem and its two extensions belong to a different family of problems, often called "Cartesian product problems," which involve finding all possible ways of combining the elements of different sets. In the Outfits and Extended Outfits Problems we have two such sets, the set of dresses and the set of hats, while in the Further Extended Outfits Problem we also have the set of socks.

As in the other families of problems discussed in Episode E, there is an efficient method for enumerating all possibilities in Cartesian product problems.[5] Yet, the specific Cartesian

[5] The cardinality of the set returned by the Cartesian product of two or more sets is equal to the product of the cardinalities of all individual sets.

	dress key				
1	OD orange dress	1. OD+GH+WS	1 2 7	1 5 7	1 4 7
2	GH green Hat	2. OD+RH+YS	1 2 8	1 5 8	1 4 8
3	GD green dress	3. OD+PH+GS	1 2 9	1 5 9	1 4 9
4	RH red Hat	4. GD+GH+WS	1		
5	PH purple Hat	5. GD+RH+GS	3 2 7	3 '	
6	BD black dress	6. GD+PH+YS	3 2 8		
7	WS White Socks	7. BD+GH+GS	3 2 9		
8	GS Green Socks	8. BD+RH+YS			
9	YS yellow socks	9. BD+PH+WS			

Figure 6.17 A student's work on the Further Extended Outfits Problem illustrating the shift from a code based on letters to a code based on digits.

product problems used by Howard in Episode F, which illustrate problems that are rather common in elementary school mathematics (save for the expectation for a proof), do not require knowledge or use of efficient enumeration methods for their solution. The relatively small number of possibilities involved in these problems makes them accessible to elementary students who can solve them using the mode of argumentation associated with the systematic enumeration of all possibilities. In this sense, proving tasks that belong to the family of Cartesian product problems can be considered to be structurally similar to proving tasks that belong to the families of problems discussed in Episode E, and thus may be expected to generate similar proving activity in the classroom.

Indeed, similar to the proving tasks implemented by Ball in Episode E, the proving tasks implemented by Howard in Episode F generated proving activity that revolved around the mode of argumentation associated with the systematic enumeration of all possible cases involved in a situation. Episode F presented a wealth of students' solutions to the various problems that used, or attempted to use, this mode of argumentation. A factor that makes it difficult to draw any definite conclusions about students' use of this mode of argumentation is that they rarely stated explicitly in their written work the systematic rule that they followed, or attempted to follow, in listing the different possibilities. However, the systematicity that characterized the lists of many students allows little if any doubt that these students generated those lists on the basis of specific rules. One can infer quite securely what these rules were by inspecting students' work.

Consider, for example, Irene's solution to the Outfits Problem (Figure 6.4). The unstated, albeit easily inferred, rule that Irene used to draw all the possibilities was the following: Begin with all possible outfits using an orange dress; there are only two such outfits as there are only two different hats (one green and one red) that can go with any given dress. Continue with all possible outfits using a green dress; there are only two such outfits for the same reason as before. Finish with all possible outfits using a black dress, which is the only remaining dress; again, there are only two such outfits. The systematicity characterizing Irene's drawings is further reflected in the fact that, for each color dress, she drew the two

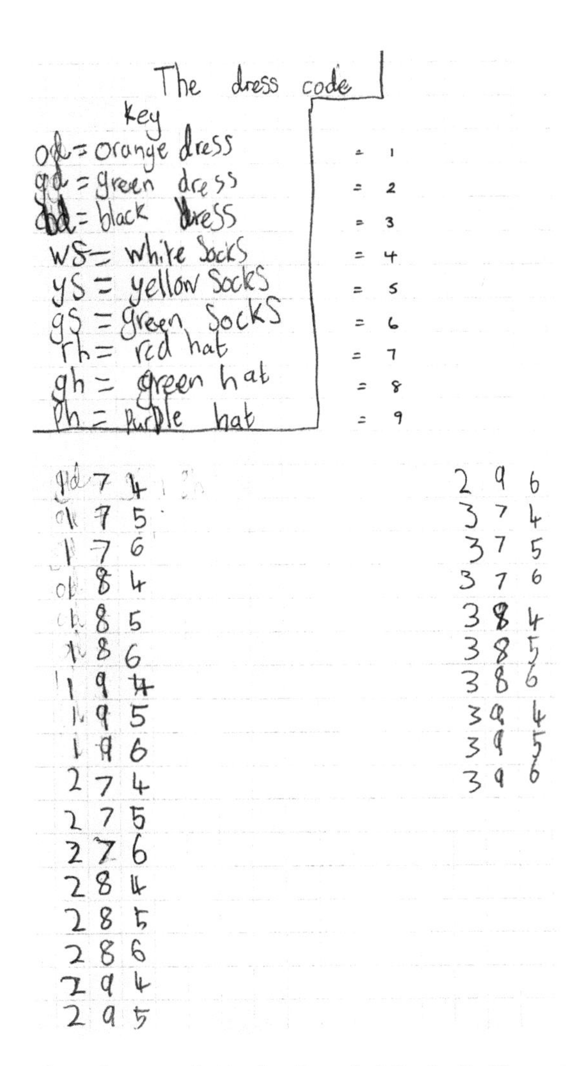

Figure 6.18 A complete solution to the Further Extended Outfits Problem using a code based on digits.

possible outfits in the same order: first, the specific dress with the green hat; second, the same dress with the red hat. A similar rule can be inferred by inspection of Jim and Andy's solution to the Outfits Problem using multi-link blocks (Figure 6.5a) and Helen's solution to the Extended Outfits Problem using drawings (Figure 6.7).

The orderly presentation of all the possible outfits in these solutions makes it highly unlikely that these students generated their lists in a random or an unmethodical way. The extra complexity that was introduced with the inclusion of different color socks in the outfits of the Further Extended Outfits Problem eliminates any reasonable doubt about whether the students who produced solutions such as those in Figures 6.8(a) and 6.18 followed a rule. The same applies for the solutions to the first and second parts of the Frogs Problem that are presented in Figures 6.12 and 6.14 (the middle and last columns), respectively.

One may conclude, then, that the students who generated all of the aforementioned solutions (Figures 6.4, 6.5a, 6.7, 6.8a, 6.12, 6.14, and 6.18) used the valid mode of argumentation associated with enumerating systematically all possibilities involved in a situation. More generally, one can say that the arguments that were reflected in these solutions satisfied all three criteria for an argument to meet the standard of proof, as described in Chapter 2:

(1) They used (implicitly or explicitly) true statements that were readily accepted by the class, such as there were only two outfits one could make using a specific dress and two different hats (see, e.g., Irene's solution to the Outfits Problem; Figure 6.4).
(2) They used a valid mode of argumentation as discussed earlier.
(3) They were represented appropriately using written language (notably drawings) or concrete materials (multi-link blocks), and in a way that the students in the class could easily understand.

This episode also presented several solutions by students in which the generated lists were incomplete (e.g., Figures 6.11 and 6.15–6.17), or, even if they were complete, they did not seem to have been generated by consistently following a specific rule (e.g., Figure 6.13). While the arguments reflected in these solutions do not meet the standard of proof, as they fail to satisfy the criterion for a valid mode argumentation, one cannot call them "empirical." This is primarily because there is no evidence to suggest that the students who produced these solutions considered that they (justified they had) found all the possible cases. In fact I hypothesize that these students considered their solutions to be "work in progress" rather than finished arguments that they believed met the standard of proof (cf. Stylianides & Stylianides, 2009a).

This hypothesis finds support from the written work of several students in this episode. For example, Irene produced an incomplete list (the left column in Figure 6.14) before she moved on to produce a systematic list (the middle and right columns in the same figure), presumably benefitting from further independent thinking about the problem or input from a peer. Also, some students attempted to use a valid mode of argumentation and began to craft a rule for systematization, which, however, they were unable to sustain due to the complexity of their chosen mode of representation (Figures 6.16 and 6.17). The available evidence in students' written work suggests that these students were unsatisfied with their original solution path (using a code based on letters) and started to explore an alternative path (using a code based on digits).

Thus, unlike Ball in Episode E, Howard did not face challenges with her students offering empirical arguments that she would then have to help them recognize as invalid. However, Howard faced a different challenge, namely, to help her students see an "intellectual need" (Harel, 1998) for, and learn how to use, efficient modes of representation in enumerating all possibilities involved in a situation. Howard seemed to have anticipated this challenge and tried to address it in different ways in her teaching during the episode.

Specifically, she started the lesson by introducing the students to different modes of representation they could use in solving the first proving task (line 2), and she allowed time for students to experience at first hand the slow progress associated with detailed or realistic illustrations of the different possibilities. Furthermore, she engaged the class in an explicit discussion of this issue (lines 6–14), thus helping increase "awareness" (e.g., Mason, 1998; Stylianides & Stylianides, 2014b) among the students of the role played by the choice of a mode of representation in their work. In the discussion students themselves commented on the impractical nature of detailed or realistic illustrations, beginning to recognize the importance of more efficient modes of representation. Howard revisited and reinforced the importance of this issue several times during the episode, also supporting her students' choice and use of modes of representation during their work on the different tasks.

In particular, in the work of the class on the Outfits and Extended Outfits Problems, Howard did the following: she highlighted the simplicity of representation in different students' solutions that were based on concrete materials or drawings (e.g., Figures 6.4 and 6.5a); she proposed ways of further simplifying some of these representations (Figure 6.5b); and she engaged the class in reflection on the emerging issues (lines 18–21). The class recognized the need for efficient ways of representing the possibilities in the different tasks. This is reflected in the work of those students who originally used complex representations but then shifted to simpler ones; compare, for example, Samantha's initial drawing of an outfit in Figure 6.3 and her later drawing in Figure 6.8(b).

In the work of the class on the Further Extended Outfits Problem, Howard took advantage of the increased representational complexity of the specific problem to again raise the issue of finding an efficient way to represent the different possibilities. She introduced to the class the idea of a "code" (or "key"), similar to a map key in geography, and she engaged the students in discussion of what that code might be if one used letters or digits (lines 33–43). She also asked the students to solve again the Further Extended Outfits Problem using their own code.[6] As I discussed earlier, those students who attempted to use a code based on letters encountered many difficulties due to the representational complexities deriving from their chosen code. These difficulties appeared to motivate several of these students to seek a new code based on digits, which indeed eased tremendously the systematic generation of all possibilities.

Finally, although the purpose of the proving tasks used in the episode was that of *justification,* the statements to be justified were not given to the students and this resulted in a couple of instances students formulating originally false statements, which then created a need for *refutation*. For example, some students originally hypothesized that the answer to the Further Extended Outfits Problem would be 12 or 18, instead of the correct answer of 27, but then they refuted their original hypotheses by justifying the correct answer (lines 26–32). Also, although the proving tasks in the episode were intended to have clearly stated conditions, there were a couple of points in those conditions that, in the eyes of some students, were subject to different assumptions. See, for example, Irene's question (line 4)

[6] The students also had the option to work on a new problem, the Frogs Problem. Interestingly, the students who opted to work on the Frogs Problem also used a code in their solutions, even though they were not explicitly instructed by the teacher to do that.

about whether an outfit in the Outfits Problem could include a dress but not a hat, or Cameron's beginning work on the Frogs Problem (Figure 6.10) based on the assumption that the lily pads were of different colors. Howard made the decision not to explore with her students the implications of solving these tasks based on different assumptions, choosing to simply clarify for the students the interpretations she had intended for the tasks. This decision is consistent with my discussion in Chapter 4 concerning the notion of "learning residue" (Davis, 1992) as one of the factors that can inform teachers' decisions about when to pursue the implications of different assumptions for solving a proving task: If Howard had chosen to pursue the implications of different assumptions for solving the two tasks, the attention of the class would have deviated to a different issue and, as a result, the students could have been deprived of the important learning opportunities to reflect on the choice of a mode of representation in listing all the possibilities involved in a situation.

General Discussion

Episodes E and F offered complementary views on what it might mean or look like when elementary teachers engage their students with proving tasks that involve multiple but finitely many cases. In this final section I discuss some general issues about the relationship between proving tasks and proving activity, in the particular context of proving tasks that involve multiple but finitely many cases, and about the role of the teacher while implementing this kind of task in the classroom.

The Relationship between Proving Tasks and Proving Activity

The teachers in these two episodes implemented in their classes a number of tasks that, strictly speaking, belonged to different families of problems: combination problems with repetition; permutation problems without repetition; and Cartesian product problems. Notwithstanding these differences, however, all of the tasks had a similar mathematical structure and a strong potential to engage students, as they did, in similar proving activity: Each task asked students to find all possibilities involved in a situation (multiple but finitely many) and justify that all possibilities have indeed been found. The proving activity that was generated by the tasks revolved around a specific mode of argumentation, namely, enumerating systematically all possibilities involved in a situation, which is important not least because of its application in proof by exhaustion. While there are efficient methods for enumerating all possibilities in the families of problems to which the tasks belonged, the relatively small number of possibilities that were involved in the tasks made the tasks accessible to elementary students and solvable using the aforementioned mode of argumentation without requiring knowledge of efficient enumeration methods.

The fact that the proving activity in the two episodes was similar across tasks and revolved around the same mode of argumentation offers illustrative evidence for a relationship between the particular kind of proving tasks and the proving activity that the implementation of these tasks can potentially generate in an elementary classroom. Having said that, a closer examination of the proving activities in the two episodes highlights two important issues that played out differently in the two classes. In what follows I describe the issues and try to interpret the observed differences in the two classes.

The first issue relates to students' tendency, observed in Episode E, to offer empirical arguments, claiming that they have found all possibilities on the basis of unmethodically created lists and seeing little "intellectual need" (Harel, 1998) to craft a rule for systematization. The episode happened at the beginning of the school year and involved Ball's first efforts to introduce her third graders to the idea of proof. Given how difficult the concept of proof is even for advanced secondary students and adults (e.g., Harel & Sowder, 2007; Küchemann & Hoyles, 2001–2003; Morris, 2007), with many of them considering empirical arguments to be proofs, the work of Ball's students in Episode E should be unsurprising. But why was the same issue not observed in Episode F?

A possible reason is that Howard had already introduced her students to the importance of listing systematically all possibilities involved in a situation as a way to justify that all possibilities have been found (see, in particular, Episode A in Chapter 4). This reason is consistent with an argument I discussed in Chapter 2, that elementary students are able to engage successfully with proof so long as they are offered appropriate learning experiences in supportive classroom environments. However, this reason falls short of explaining the apparent contradiction between the relative comfort with which Howard's students sought to produce non-empirical arguments and the persistent difficulties that older students and adults tend to have in recognizing the limitations of empirical arguments.

A possible resolution of this contradiction is the following: The research that reported difficulties of older students and adults with empirical arguments tended to examine those difficulties in the context of proving tasks involving *infinitely* many cases, which are quite different and presumably more difficult than the proving tasks used by Howard in Episode F. Indeed, the mode of argumentation associated with systematically enumerating all possibilities involved in a situation, which Howard's students used in their proofs in Episode F, would generally be unhelpful in proofs of mathematical generalizations over infinite sets. In fact any attempts by solvers to use this mode of argumentation in proofs of generalizations over infinite sets would be likely to end up producing an empirical argument. Research on the understanding of proof by students and adults has paid little attention thus far to proving tasks involving finitely many cases (Stylianides et al., 2016b). This is problematic, for notions such as "empirical argument" take a rather different form in the context of different kinds of proving tasks, involving infinitely or finitely many cases, and so it is inappropriate to uncritically extrapolate research findings from one kind of proving task to the other.

The second issue that played out differently in the proving activities described in the two episodes relates to the role of the mode of argument representation in students' solutions. On the one hand, in Episode F the choice of representational tools featured prominently in the work of the class, being the explicit focus of discussion in many instances during the episode and appearing to influence students' success in crafting a rule for systematization to solve some of the proving tasks (notably the Further Extended Outfits Problem). On the other hand, in Episode E the role of the mode of argument representation was tacit in students' solutions, staying in the background of the mathematical work of the class. How can we explain this difference?

A reason for it may be found in the particular formulation of the main proving tasks in the two episodes: While the Date Problem in Episode E described a situation involving simple representations (three digits) that students could directly take and use in listing different possibilities, the Outfits Problem and its extensions in Episode F tempted students into using complex representations (notably realistic drawings) that ended up hindering students' efforts to list all the possibilities. Indeed, in my discussion of Episode F I highlighted the catalytic role that students' shift toward more efficient modes of argument representation played in their ability to solve the Further Extended Outfits Problem: from using realistic drawings of outfits; to using a code based on letters; to using a new code based on digits.

Thus, the comparison of students' proving activity in the two episodes suggests that two structurally similar proving tasks that call for use of the same mode of argumentation can nevertheless place significantly different cognitive demands on students. As illustrated by these episodes, this can happen when the situation described in one of the tasks gives to students a simple and ready-to-use representation, while the situation in the other task tempts students into using a complex and inefficient representation. Interestingly, despite wide appreciation of the importance of representations and mathematical language more broadly as tools for communication linked to students' mathematical understanding (e.g., Lamon, 2001; Pyke, 2003; Sfard, 2001), the modes of representation in students' argument constructions have been the explicit focus of relatively few studies thus far. These studies have offered useful insights into the role of diagrams in the argument constructions of undergraduate students or mathematicians (e.g., Alcock & Simpson, 2004; Gibson, 1998; Samkoff, Lai, & Weber, 2012) and the use of concrete materials or pictures as a way to reduce abstraction and support young children in constructing arguments that approximate the standard of proof (e.g., Maher et al., 2010; Morris, 2009; Schifter, 2009). More research is needed to understand the relationship between the specific formulation of related proving tasks and the cognitive demands that these tasks place on (elementary) students' proving activity, especially with regard to students' choice and use of different modes of argument representation.

The Role of the Teacher

Episodes E and F offer some insight into the role of a teacher who implements a proving task that asks students to find all possibilities involved in a situation (multiple but finitely many) and justify that all possibilities have indeed been found. This relates to the teacher being able to anticipate, and prepared to address, certain difficulties that students might encounter as they engage with this particular kind of proving task. While this aspect of a teacher's role is not specific to the area of proof (e.g., Stein et al., 2008), the two episodes illustrate, in complementary ways, what this aspect might entail for elementary teachers during the classroom implementation of this particular kind of proving task.

Episode E showed that elementary students, especially those who are just being introduced to the concept of proof, are likely to produce empirical arguments, with little appreciation of the importance of crafting a rule for systematization to show that they have found all possibilities involved in a situation. Ball had anticipated and aimed to address this difficulty, primarily by playing the role of the "skeptic" (Mason, 1982): She showed dissatisfaction with students' empirical arguments and pointed out to them that one would have to make a leap of faith in accepting their claim that all possibilities had been found. Ball also gave a lot of discussion time to Betsy's argument, which, as I explained earlier, met the standard of proof. By so doing, she offered the class an opportunity to see, and begin to appreciate, what a viable alternative to empirical arguments might be or look like.

The challenge faced by Ball about how to create a need for proof among her students is an important but stubborn problem in mathematics education (e.g., Zaslavsky, Nickerson, Stylianides, Kidron, & Winicki, 2012). While some research has reported promising ways of helping secondary students or adults begin to recognize the limitations of empirical arguments and see an "intellectual need" (Harel, 1998) for more secure validation methods (i.e., proofs) (Stylianides, 2009a; Stylianides & Stylianides, 2009b), this research was carried out in the context of proving tasks that involved an infinite number of cases, and thus it addressed forms of empirical arguments that are encountered in that context. However, as I discussed earlier, research findings related to one kind of proving task may not be readily applicable to another. Similarly, research findings with secondary students or adults cannot securely be extrapolated to elementary students' engagement in proving.

Episode F showed that, even if elementary students have a sufficient understanding of the importance of a rule for systematization in solving the particular kind of proving task, students might still have difficulty in choosing and using an efficient mode of representation to list all possibilities. As I discussed earlier, Howard had anticipated and aimed to address this student difficulty in a number of ways, including the following: She allowed space for students to experiment with and begin to experience for themselves the inefficiency of certain modes of representation; she organized classroom discussions that offered a forum for critique of and reflection on different modes of representation; she modeled for students the use of efficient modes of representation and directed their attention to specific arguments that used efficient modes; and she gradually transferred responsibility to students to develop and apply an efficient mode in the context of a given problem. These ways relate to some pedagogical practices discussed in the literature, such as: strategies for orchestrating productive classroom discussions (Stein et al., 2008; Stylianides & Stylianides, 2014b); approaches to bringing to the students' attention key mathematical ideas or heuristics that might have already been embedded in their work (Mason, 2009); and "scaffolding" students' work by adapting instruction to respond to their difficulties and gradually withdrawing the offered support, thus facilitating handover to independence (Bakker, Smit, & Wegerif, 2015).

Finally, Episode F suggests the general structure of a possible "learning trajectory" (Simon, 1995) that teachers might aim to support when they engage their students with proving tasks, such as the different versions of the Outfits Problem, that tempt students into using complex and inefficient representations. This learning trajectory can comprise the following points: (1) trying to use the complex mode of representation they are enticed into by the tasks and recognizing its inefficiency; (2) seeking to develop an alternative, more efficient, mode of representation; and (3) developing an argument for the tasks using an efficient mode of representation. The transition between the first two points is crucial, and unless it is recognized as important and deliberately supported by the teacher, as Howard did in Episode F, students' work runs the risk of being mathematically unproductive, with lesson time being spent more on non-mathematical features of the tasks (e.g., portraying realistically the different outfits) and less on mathematical reasoning (cf. Stylianides & Stylianides, 2008a). In describing and discussing Episode F, I offered samples of student work illustrating the different points of the outlined learning trajectory. However, research is needed for a better understanding and more nuanced description of these points and of their connections, including the different student arguments (both valid and invalid) one might expect to observe in students' work as they progress through the learning trajectory. While many categorizations of different student arguments have been proposed and used in the literature (e.g., Balacheff, 1988a; Harel & Sowder, 2007; Marrades & Gutiérrez, 2000; Simon & Blume, 1996), these categorizations are not readily applicable for addressing the specific issue raised here as they are more tailored to a different kind of proving tasks (namely, proving tasks involving an infinite number of cases).

7

Proving Tasks Involving Infinitely Many Cases

In this chapter I examine proving tasks involving infinitely many cases, with an emphasis on the proving activity that this kind of task can help generate in the classroom and on the role of the teacher while implementing the tasks. In Chapter 4 I discussed several examples of this kind of task, which involved justification that different sets had an infinite cardinality. The focus here is not on the cardinality of the infinite sets involved, but rather on generalizations formulated over such sets.

As in Chapters 4–6, I situate my discussion in the context of two classroom episodes: the first (Episode G) comes from Howard's Year 4 class in England (8–9-year-olds); the second (Episode H) comes from Ball's third-grade class in the United States (again 8–9-year-olds). These episodes have been selected for their illustrative power and for the complementary issues they raise about what it might mean or look like when elementary teachers use in their classes this particular kind of proving task. I describe and discuss each episode separately, and I conclude with a more general discussion.

Episode G

Description of the Episode

This episode took place in Howard's Year 4 class and describes work that happened during two lessons, on June 10 and 17. The second lesson was scheduled a week after the first so as to accommodate my schedule restrictions and enable my presence in the class.

The lesson on June 10 began with Howard reminding the students of the following conjecture, which they had formulated at the end of the previous lesson as part of their work on the ratio and proportion curriculum unit. I call this Conjecture 1 for ease of reference.

1. *Conjecture 1*: "I think when you multiply two numbers together, the answer gets bigger."

Following a brief recap of what the class meant by "conjecture," Howard gave the students an opportunity to further explore Conjecture 1:

2. *Howard*: This is our conjecture from yesterday. What does "conjecture" mean?
3. *A girl*: Something we think is true.
4. *Howard*: Take some time to think a bit more about the conjecture. Choose a partner to work with.

I was sitting close to a table with Ryan, Daniel, and the teaching assistant of the class (whom I have not seen in previous lessons). Ryan had already written on his paper a couple of confirming examples when the teaching assistant asked:

5. *Teaching assistant*: Can you think of a case when this [Conjecture 1] is not true?
6. *Ryan*: [After doing some thinking:] If you multiply by zero, the answer doesn't get bigger. [He then wrote on his paper the product, $10000 \times 0 = 0$, and said:] Even if you multiply a trillion by zero, the answer would still be zero, not bigger!

A few minutes later Ryan noticed that a similar situation applies for multiplication by the number 1:

7. *Ryan*: If you multiply by 1, the answer doesn't get bigger either. It stays the same.

In the whole class discussion that followed, Pam and Emma presented $1 \times 1 = 1$ as an example of a product where the answer does not get bigger. Howard asked whether other students had thought of multiplying by the number 1; about six students raised their hands. Then Cameron talked about multiplying by zero, presenting similar work as Ryan had done earlier during small-group work (line 6). Following a question from Howard, about half of the students in the class confirmed that they, too, had thought about multiplying by zero.

8. *Howard*: What happens when we check more examples?
9. *Simon*: We get more evidence.
10. *Howard*: How many examples do we need to show that this conjecture [Conjecture 1] is not true?
11. *Daniel*: Two.
12. *Andy*: Four.
13. *Orrin*: Fifteen.
14. *Howard*: Okay, we'll come back to this.

After some other student contributions, a boy mentioned a few examples where multiplication made bigger, such as $2 \times 3 = 6$, and said:

15. *A boy*: I believe the conjecture works.
16. *Howard*: But what about all these counterexamples we mentioned earlier? [The counterexamples were also written on the board and Howard pointed at them.]

17. *The same boy*: [He stayed silent but his facial expression suggested that he was puzzled over Howard's comment.]

The discussion then shifted to multiplication with negative numbers:

18. *Gina*: We looked at minus numbers.
19. *Cameron*: Minus one times minus one goes up to zero. [Howard wrote on the board what Cameron said, but in the meantime Cameron changed his mind:] I have changed my mind.
20. *Howard*: It's okay, we'll leave this for now. We'll work only with one minus number and a positive number.
21. *A boy*: Five lots of minus four makes minus twenty. [Howard wrote on the board: $5 \times -4 = -20$.]
22. *Howard*: Where would the one lot of minus four go on the number line? [She drew a vertical number line with the position of zero marked on it.]

The boy showed the correct position of –4 on the number line. Later a girl tried to show the position of "two lots of –4," but she marked –2 and said that the answer would be halfway between –4 and 0. Another girl showed the correct position of "two lots of –4" (i.e., on –8), and she explained: "It has to be below –4 as the number gets smaller." Howard then led the discussion to a closure, saying that the class would revisit negative numbers in a future lesson.

Soon after that, two students offered two different number sentences involving decimal numbers:

23. *The number sentence offered by a girl*: $5 \times 0.1 = 0.5$
24. *The number sentence offered by Andy*: $5 \times 0.2 = 1$

Howard pointed at Conjecture 1 and asked the students whether they still believed it was true. The vast majority of the class said they believed it was not true; two students said they thought it was true; and a few others expressed no opinion. One girl proposed generating some examples involving fractions and Howard asked the class to take a few minutes to do that. In the meantime, Howard went to talk with Fiona, one of the two students who said that Conjecture 1 was true.

25. *Howard*: So you think the conjecture is true?
26. *Fiona*: It's *sometimes* true: it works in some examples but not in others. [Fiona then mentioned some specific examples.]

Having observed this exchange between Howard and Fiona, I suggested to Howard that she rephrased the conjecture as follows so as to make explicit the generality of the statement. I call this re-formulation Conjecture 1a:

27. *Conjecture 1a*: "I think when you multiply two numbers together, the answer *always* gets bigger."

This was the only time during my classroom observations in Howard's class when I intervened with a specific suggestion that Howard could implement in real time. I felt that the specific classroom situation offered an opportunity to explore the possible role of language in Fiona's and possibly other students' difficulties in recognizing that the conjecture was false. Howard followed my suggestion and rephrased the conjecture as in line 27. She announced the new phrasing to the class, emphasizing the insertion of the word "always," and she asked the students to consider this phrasing as they tried to generate examples involving fractions.

At the table where Daniel, Ryan, and the teaching assistant were sitting, the teaching assistant invited the two students to say what they thought about Conjecture 1a.

28. *Daniel*: I think it's sometimes true, sometimes false.
29. *Ryan*: I think it's always true except from these [he pointed at the counterexamples he had written on his paper].
30. *Teaching assistant*: So is it *always* true?
31. *Ryan*: Not *always*!

In the whole class discussion that followed different students used fraction strips to present number sentences such as the following.

32. *Harriet*: One fourth of 16 is 4.
33. *Irene*: One third of 6 is 2.

With questioning from the teacher, the students indicated that the word "of" in these number sentences meant multiplication. The class agreed also that in these examples "multiplication made smaller." Howard then asked:

34. *Howard*: What if instead of "always" [in the statement of Conjecture 1a] we had "sometimes"? Would that be true?

With these questions Howard essentially introduced a new re-formulation of the conjecture, which I call Conjecture 1b:

35. *Conjecture 1b*: "I think when you multiply two numbers together, the answer *sometimes* gets bigger."

All students in the class raised their hands to indicate their agreement that Conjecture 1b was true.

Howard then challenged the students to think how they could complete the following sentence, which is essentially an extended version of Conjecture 1a. I refer to it as (the unfinished statement of) Conjecture 1c:

36. *Conjecture 1c*: "When you multiply two numbers together, the answer always gets bigger if. . . ."

1	2	3	4	5	6	7	8	9	10
11	12	13	14	15	16	17	18	19	20
21	22	23	24	25	26	27	28	29	30
31	32	33	34	35	36	37	38	39	40
41	42	43	44	45	46	47	48	49	50
51	52	53	54	55	56	57	58	59	60
61	62	63	64	65	66	67	68	69	70
71	72	73	74	75	76	77	78	79	80
81	82	83	84	85	86	87	88	89	90
91	92	93	94	95	96	97	98	99	100

Figure 7.1 A 100-number grid.

The following ideas were offered from students about how the statement of the conjecture could be finished:

37. *Gina*: If they are in the 2-times table and beyond.
38. *Zenia*: If the number you are multiplying by is above zero. . . . [After some thinking she corrected herself:] If it is above 1.
39. *Osborn*: If the numbers are not minus and above 1.

Howard led the discussion to a closure and introduced a new conjecture involving consecutive (natural) numbers. She presented a slide that was titled "Three consecutive numbers" and contained a 100-number grid, as in Figure 7.1.[1]

40. *Howard*: What does "consecutive numbers" mean?
41. *Class*: [No student volunteers to speak.]
42. *Howard*: It means three numbers that are next to each other [on the 100-number grid] like 31, 32, 33.

After mentioning a few more examples, Howard told the class what she had discovered the night before as she was exploring sums of three consecutive numbers.

43. *Howard*: I looked at 3, 4, 5. Their sum is 12 and I said, "Oh, it belongs to the 4-times table!" I also looked at 7, 8, 9. Their sum is 24, again in the 4-times table! So that's my conjecture: "When you add three consecutive numbers, the answer is always in the 4-times table." What do you think?

[1] The use of the 100-number grid made it clear that the domain of reference was the set of natural numbers.

Andy is eager to talk, but Howard asks him to allow all students an opportunity to consider the conjecture in their small groups. A statement of the conjecture, which I call Conjecture 2, was written on the board as follows:

44. *Conjecture 2*: "I think that if I add three consecutive numbers together, the answer is always in the 4-times table."

During small-group work Howard went to speak with Andy. Andy said that the answers in the two examples that Howard mentioned earlier (line 43) were in the 4-times table, because of the middle numbers in those examples (i.e., 4 and 8), which were also in the 4-times table. Andy and Keith, who were sitting next to each other, conjectured further that the answer would always be in the times table of the middle number. Keith said to me when I visited their table:

45. *Keith*: Take 8, 9, 10. They add up to 27, which is in the 9-times table, but not in the 4-times table.

Keith mentioned a couple of other examples that confirmed their conjecture.

In the whole class discussion that followed, a boy offered $1+2+3=6$ as an example of a sum that does not belong to the 4-times table. The boy concluded that Conjecture 2 was not true.

46. *Howard*: How many examples do we need to show that it [Conjecture 2] is not true?
47. *Irene*: Quite a lot!
48. *A boy*: About ten.
49. *Harriet*: Five.
50. *Andy*: Three. Not very many.
51. *Howard*: How many exactly?
52. *Andy*: Between three and five.
53. *Irene*: Some [examples] are in the 3-times table and some are in the 4-times table.
54. *A girl*: It's definitely not always true. It's linked to the middle number.
55. *Andy*: It's in the times table of the middle number.
56. *Another boy*: If the middle number is in the 4-times table, then the answer is going to be there too.

Howard asked the students to think how they could revise the conjecture by completing the following sentence, which I refer to as (the unfinished statement of) Conjecture 2a:

57. *Conjecture 2a*: "I think that if I add three consecutive numbers together, the answer is always in. . . ."

I noted earlier that Andy and Keith had already come up with a way to complete this sentence, but their work was not presented publicly. As a context for students' work on Conjecture 2a, Howard wrote on the board the following examples that she got from the paper of two girls:

58. *List of examples generated by two girls*:

$$1+2+3=6$$
$$2+3+4=9$$
$$3+4+5=12$$
$$4+5+6=15$$
$$5+6+7=18$$

Pam initiated the following whole class discussion:

59. *Pam*: All the numbers are in the 3-times table.
60. *Howard*: Isn't it interesting? Maths is all about patterns!
61. *A boy*: Some numbers though are in the 6-times table.
62. *Howard*: Why do you think that is?
63. *The same boy*: [After he did some thinking:] Because the 6-times table is in the 3-times table.
64. *Howard*: Nice thought. Which other number's the times table is in the 3-times table?
65. *The same boy*: Nine.
66. *Howard*: So there is a link between these times tables.

This was the end of the lesson on June 10. The lesson on June 17 started with a recap of students' work from the previous week. Howard reminded the class that they had been trying to complete the statement of Conjecture 2a (line 57) by considering also some examples (line 58), and that they had observed all the answers were in the 3-times table.

67. *Daniel*: Nine is in the 3-times table but 3 is not in the 9-times table.
68. *Howard*: Maybe it's better if we don't use [the word] "times table." We can use instead another word we've learned recently. What's that word?
69. *A girl*: [With some hesitation:] Multiple.
70. *Another girl*: They are all multiples of 3.
71. *Howard*: Talk with a partner: What do you understand by "multiple of 3"?
72. *Eve*: It's in the 3-times table.

The students spent some time discussing the meaning of a multiple of 3 and generated several examples and non-examples of a multiple of 3. The class determined whether a given number was a multiple of 3 either by finding a whole number which when multiplied to 3 gave that number or by dividing that number by 3 and seeing whether the result was a whole number. One of the numbers checked by the class was 509. Howard took a calculator and asked the students to say what kind of number they would expect to get when they divided 509 by 3 if 509 was a multiple of 3. Irene said "a number without a decimal point." Howard did the division using the calculator and announced the result to the class: a decimal number. The class concluded that 509 was not a multiple of 3.

The class then worked on the following conjecture, which was a completed version of the statement of Conjecture 2a:

73. *Conjecture 2b*: "I think that if I add three consecutive numbers together, the answer is always a multiple of 3."

During small-group work Helen and Rania suggested to Howard a re-formulation of this conjecture based on the observation that the result of the addition is three times the middle addend. Andy and Keith had also been working on this version of the conjecture, which can be phrased as follows.

74. *Conjecture 2c*: "I think that if I add three consecutive numbers together, the answer is always 3 times the middle number."

Howard shared this conjecture with the rest of the class. Different students recognized the link between the statements of Conjectures 2b and 2c. For example, at two different points during the whole class discussion Harriet and Andy said the following:

75. *Harriet*: If you times the middle number by 3 [i.e., the result described in Conjecture 2c] you get a multiple of 3 [i.e., the result described in Conjecture 2b].
76. *Andy*: Multiply the middle number by 3 equals a multiple of 3.

Howard invited the students to use multi-link blocks to explore Conjecture 2c and to see also whether their work from past lessons with function machines could be useful in any way. Indeed, I had already observed a couple of students trying to use algebraic notation in their work on Conjectures 2b and 2c.

77. *Howard*: It would be useful to use letters like we did with function machines. Which number would you call n?
78. *Melanie*: We called n the middle number.
79. *Howard*: Who else did that?
80. *Class*: [Several students raised their hands.]

After some time working in small groups, Howard called the class back together to present to them one idea that Jack had just explained to her. Jack worked on the particular case $4+5+6$, which he modeled using multi-link blocks as in Figure 7.2(a). According to a discussion I had with Howard after the lesson, Jack used the particular case $4+5+6$ as a context to explain to her more generally that, irrespective of the specific choice of three consecutive natural numbers, one would always be able to move one block from the largest number to the smallest number, as in Figure 7.2(b), thus ending up with three numbers that are all equal to the middle number.

81. *Howard*: Jack says this [she showed Jack's model as in Figure 7.2a] is three times the middle number. Why is that? Jack?
82. *Jack*: Because one number is smaller and the other is larger [than the middle number].
83. *Howard*: By how much?
84. *Jack*: One smaller and one bigger.
85. *Howard*: And then what?
86. *Jack*: [He took one block from the larger number and added it to the smaller number as in Figure 7.2b.]
87. *Howard*: Right. We can move one block from the largest number to the smallest. And then what?
88. *A boy*: They are all equal to the middle number.
89. *Howard*: If I call the middle number *n*, what would I call the smallest number?
90. *Melanie*: [You can call it] *m*.
91. *Howard*: I could do that, but how else could I call it?
92. *Another girl*: [You can call it] *n* minus 1.

The class got excited with the use of letters and observed that the largest number could be called *n* plus 1. A boy was heard to say:

93. *A boy*: This will always work!

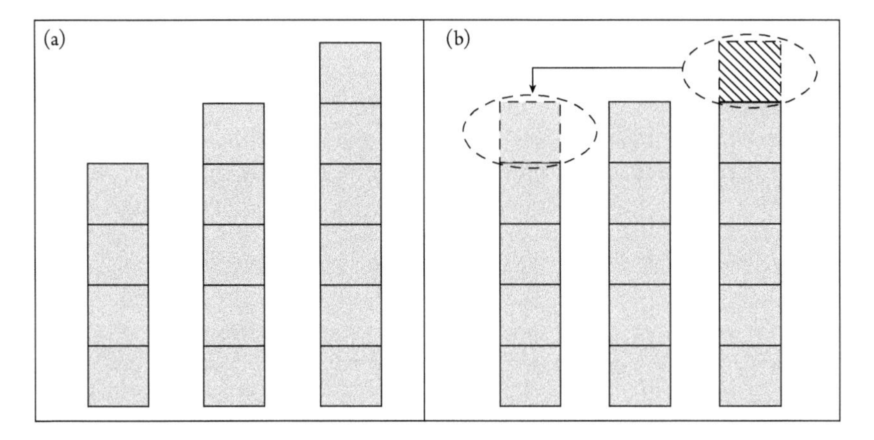

Figure 7.2 Jack's idea of moving a block from the largest number to the smallest number, illustrated in the particular case 4 + 5 + 6.

Howard asked the class to look back to Conjecture 2c and see what their work with letters meant in that context.

94. *Andy*: The answer is going to be three times *n*.
95. *Howard*: The answer when you do what?
96. *Andy*: Adding.
97. *Howard*: Adding what?
98. *Andy*: Three consecutive numbers.
99. *Howard*: Yes. And we called the middle number *n*.

Howard praised the students for their work and introduced a new task related to the concept of ratio.

Discussion of the Episode

In this episode Howard engaged the class in exploring conjectures that involved an *infinite* number of cases. The implied proving tasks for these explorations were for students to *justify* or *refute* the respective conjectures. This expectation for justification or refutation is consistent with the very meaning of a *conjecture* as a statement for which there is some doubt about its truth and thus further action is needed before it can be accepted or rejected (e.g., Arzarello, Andriano, Olivero, & Robutti, 1998; Reid, 2002; Stylianides, 2009b). The class was familiar with the meaning of a conjecture (lines 2–4).

There were two main conjectures in this episode, Conjecture 1 (line 1) and Conjecture 2 (line 44), each of which was explored and further refined by the class, thus generating a collection of related conjectures: Conjectures 1a, 1b, 1c (lines 27, 35, 36), and Conjectures 2a, 2b, 2c (lines 57, 73, 74), respectively. Conjecture 1 was about whether "multiplication (always) makes bigger," a common student misconception deriving primarily from the overemphasis of the "repeated addition" meaning of multiplication in students' early multiplication work (e.g., Fischbein et al., 1985; Haylock, 2014; Tirosh, Tsamir, & Hershkovitz, 2008). Partly due to their focus on this misconception, Conjecture 1 and related conjectures served as a context not only for proving but also for deepening and expanding students' understanding of the meaning of multiplication. Conjecture 2 and related conjectures were about developing an appropriate description for the sum of three consecutive natural numbers. Similar to the previous set of conjectures, this set also offered a productive context for a broader mathematical activity that allowed students to deepen and apply their knowledge of multiples of a number and algebraic notation.

The statement of Conjecture 1 and related conjectures was ambiguous: It did not specify whether the comparison made in it was between the product and one or both of the factors. The class operated on the tacit (unstated) assumption that the comparison was with both factors, i.e., it was assumed that the statement said the product of two factors, say factors 1 and 2, was bigger than both factor 1 and factor 2. This was a sensible assumption to make that allowed examples such as $5 \times 0.2 = 1$ (line 24) and $\frac{1}{4} \times 16 = 4$ (line 32) to be viewed as counterexamples to the conjectures, in line with what the teacher had intended. This interpretation of the statement "multiplication makes bigger" is also in line with how the statement has been used or interpreted elsewhere (e.g., Haylock, 2014; Tirosh et al., 2008). Of course the ambiguity in the statement could have allowed students to make alternative interpretations of its meaning. Any such alternative interpretations might have led to opposite conclusions about whether examples such as those that I have just mentioned offered confirming or disconfirming evidence for the conjectures. As discussed in Chapter 4, the teacher would then have an important decision to make about whether to simply clarify to students the intended interpretation or to engage them in pursuing different assumptions and reflecting on the implications of that work.

The proving activity students engaged in as they explored Conjecture 1 included lots of counterexamples. Some of these counterexamples were: $10000 \times 0 = 0$ (Ryan, line 6) or another number multiplied by zero (Cameron and about half of the other students in the class); $1 \times 1 = 1$ (Pam and Emma) or another number multiplied by 1 (Ryan, line 7, and at least six other students in the class); $5 \times -4 = -20$ (line 21); $5 \times 0.1 = 0.5$ (line 23); $5 \times 0.2 = 1$ (Andy, line 24); $\frac{1}{4} \times 16 = 4$ (Harriet, line 32); and $\frac{1}{3} \times 6 = 2$ (Irene, line 33).

All of these counterexamples refuted the conjecture and met *potentially* the standard of proof because of the following points that correspond to the criteria for a proof discussed in Chapter 2.[2]

(1) They used true statements that were readily accepted by the class, notably the results of the respective calculations.

(2) They employed the valid mode of argumentation associated with refutation of a general claim by means of a counterexample, which was *arguably* within the reach of the classroom community.

(3) They were presented appropriately using a combination of verbal and written language that was understandable to the students in the class.

With regard to point (1), which concerns the classroom community's set of accepted statements, the episode presented a couple of instances where calculation errors, rooted in students' insecure knowledge of the underpinning concepts or procedures, interfered with their efforts to construct counterexamples (see, e.g., lines 18–19 relating to knowledge about multiplying negative numbers). Howard adjusted the scope of students' mathematical work accordingly (e.g., line 20) so as to keep it within mathematical territories that students could navigate comfortably (e.g., multiplying by 1, zero, decimals, or fractions) or territories students could potentially navigate after some scaffolding from her (see, e.g., the work of the class on $5 \times -4 = -20$ starting from line 21). This was a sensible way of handling the affordances and limitations of the classroom community's set of accepted statements at the given time (Stylianides, 2007b; see also Chapter 5): A possible decision to extend students' work into new and unfamiliar mathematical territories would require time for deliberate instruction of relevant concepts or procedures and could thus deviate attention away from the work on the conjecture.

With regard to point (2), which concerns the classroom community's potential understanding or use of the valid mode of argumentation associated with refutation of a general

[2] I use "potentially" due to uncertainty about whether the valid mode of argumentation required for the proof was within the reach of the classroom community at the given time. See point (2) and subsequent discussion.

claim by means of a counterexample, the episode presented a rather complex picture (and hence my use of the word "arguably" in the description of the point). On the one hand, the class seemed to be comfortable with generating a wealth of counterexamples and recognizing that these offered disconfirming evidence for Conjecture 1. On the other hand, the class was hesitant, or seemed to be unable, to conclude, on the basis of these counterexamples, that the conjecture was false, let alone to recognize that a *single* counterexample would suffice for that conclusion. For example, Simon said that more counterexamples give us "more evidence" that the conjecture is not true (lines 8–9). Also, different students had different standards about the number of counterexamples required to show that the conjecture was not true: Daniel wanted two (line 11), Andy four (line 12), and Orrin fifteen (line 13).

This episode offers some insight into different issues involved in students' understanding of, or difficulty in understanding, refutation by counterexample. After the class had generated lots of counterexamples, including some with decimal numbers (lines 23–24), more students seemed willing to accept that Conjecture 1 was not true, but, even then, two students still believed the conjecture was true. Howard's discussion with Fiona, one of these two students, was illuminating: Fiona explained to Howard that she considered the conjecture to be *sometimes* true due to the availability of "mixed examples," with some of them offering confirming evidence for the conjecture and others offering disconfirming evidence (lines 25–26). Fiona's explanation seemed to suggest that the issue could be one of language: The way Conjecture 1 was phrased might have masked, in the eyes of Fiona and other students, the generality of the statement. Howard and I tried to explore this issue by inserting the word "always" into the statement of the conjecture, thus turning it into Conjecture 1a (line 27). Yet students' difficulties persevered, as illustrated by Daniel's description of the new conjecture being "sometimes true, sometimes false" (line 28) or Ryan's description of it being "always true except from" the identified counterexamples (line 29).

Interestingly, once Howard proposed replacing the word "always" with the word "sometimes" in the statement of Conjecture 1a (line 34), thus turning it into Conjecture 1b (line 35), the class unanimously accepted it as true. Unlike Conjecture 1a, which was false and called for refutation by counterexample, Conjecture 1b was true and called for justification by using a different mode of argumentation that seemed to be known or easily accessible to the students in the class, namely, finding few confirming and few disconfirming examples for the statement of the conjecture. The apparent difference in students' response to these two related conjectures raises questions about their conceptions of adverbs of quantification such as "always" and "sometimes," as well as how these conceptions might interfere with students' ability to recognize the truth or falsity of mathematical generalizations over infinite sets. I return to this issue in the general discussion at the end of the chapter.

The proving activity that students engaged in as they explored Conjecture 2 presented lots of similarities with their exploration of Conjecture 1: Students like Keith (line 45) quickly generated counterexamples to Conjecture 2, and the issue was raised again about how many counterexamples were needed to show that the conjecture was not true. Students expressed similar ideas as before, which indicated their difficulty in recognizing that a single counterexample is sufficient to refute a mathematical generalization (lines 46–52). As the students worked with and reflected on more examples of sums of three consecutive natural numbers (e.g., lines 58–66), and after some input from the teacher regarding the relevance of multiples of a number to the work at hand (lines 67–72), the class revised Conjecture 2 into its forms 2b and 2c (lines 73 and 74, respectively). Both these new conjectures were true and, as observed also by students (lines 75–76), offered complementary ways to describe the sum of three consecutive natural numbers.

This episode presented an argument for Conjecture 2c that was based on an idea from Jack and was further developed in a whole class discussion led by Howard (lines 81–99). Based on information I got from Howard after the lesson, I conclude that the argument that Jack presented to her during their private conversation was an example of a *generic argument* (e.g., Balacheff, 1988a; Mason & Pimm, 1984; Movshovitz-Hadar, 1988; Rowland, 2002; Tall, 1999). Specifically, Jack formulated his argument in the particular case $4+5+6$, which he viewed as a representative of the whole class of sums of three consecutive natural numbers. The argument essentially went as follows:

> We start with three consecutive natural numbers, say 4, 5, and 6, which we can model using multi-link blocks (as illustrated in Figure 7.2a). If we move one block from the largest number, 6 in the particular case, to the smallest number, 4 in the particular case, then we end up with three numbers that are all equal to the middle number, 5 in the particular case (as illustrated in Figure 7.2b). So the sum of the three consecutive numbers we started with is three times the middle number, 3×5 in the particular case. But it doesn't matter what three consecutive numbers we start with: we can always move one block from the largest number to the smallest number, thus ending up with the same three numbers, all equal to the middle number. So the sum will always be three times the middle number.

As discussed in Chapter 2, generic arguments such as this one use valid modes of argumentation and satisfy the definition of proof. Indeed, the argument presented above meets the standard of proof because:

(1) It used true statements that were readily accepted by the class, notably the fact that moving one block from the largest number to the smallest number in a collection of three consecutive natural numbers results in three numbers that are all equal to the middle number.
(2) It used a valid mode of argumentation that presumably was within the reach of the classroom community and was associated with the use of accepted facts (such as the above) to deduce logically the statement of the conjecture.
(3) It was presented appropriately using a combination of concrete materials and verbal language, and in a way that was understandable to the students in the class.

Howard played a key role during the presentation of Jack's argument to the rest of the class: She asked Jack several questions that helped construct the argument step by step, also using contributions from other students in the class (lines 81–99). Furthermore, she attempted to highlight the generality of Jack's argument by inviting students to use letters to represent the three consecutive natural numbers, starting them off by calling the middle number n (line 89). The class responded fairly well to the challenge posed to them by Howard and ended up with this representation: $n-1$ for the smallest number; n for the middle number; and $n+1$

for the largest number. Andy (and possibly other students in the class) saw that the sum of these three expressions was "3 times n," linking it, after prompting from Howard, to the statement of the conjecture (lines 94–99). I presume that the way in which Andy saw the result "3 times n" was not by simplifying the expression $(n-1)+n+(n+1)$. Indeed, this simplification would require knowledge of manipulating algebraic expressions that was unavailable to the class at the given time. Rather, I presume that Andy was able to see the result "3 times n" by linking the re-representation of Jack's argument using letters to its original representation, which was based on concrete materials and was formulated in the particular case $4+5+6$.

To conclude, this episode presented rich mathematical activity, with proving as the cornerstone of this activity and as the vehicle through which students engaged in making sense of, or deepening their understanding of, concepts such as multiplication and multiples of a number. The proving activity in particular spanned a number of related conjectures that got students involved in refuting or justifying statements formulated over infinite sets. The students experienced difficulties in recognizing the power of a single counterexample to refute a false mathematical generalization and more generally in understanding what it takes to establish that a conjecture is false. At the same time, however, the students presented competence in constructing examples that offered disconfirming evidence for a false conjecture and also in revising such a conjecture so as to make it true. The teacher played a key role in managing students' mathematical activity in the episode. Specifically, she orchestrated several discussions about what is involved in refuting a conjecture and used different false conjectures as triggers for that discussion. Also, she used two separate collections of related conjectures that she sequenced carefully, thus offering students the opportunity to revisit and deepen their evolving understanding of issues pertaining to exploration, refutation/revision, and justification of mathematical generalizations. Furthermore, through prompting and skillful questioning she led the presentation to the class of a generic argument for a true generalization over an infinite set, as well as the further development of this argument to a more general proof using algebraic notation. The leading role that the teacher played during the presentation and further development of this argument reflects the challenge involved in getting students of this age to independently develop or present such an argument. At the same time, however, the teacher's actions were capitalizing on and extending specific student contributions during the lesson, an indication that the proving activity at hand was within students' reach. Also, one might not reasonably expect students to act independently in solving a challenging task without having been first offered sufficient opportunities to engage with relevant ideas and with scaffolding from the teacher (cf. Bakker et al., 2015).

Episode H[3]

Description of the Episode

This episode took place in Ball's third-grade class and contains work that happened in three consecutive classes close to the middle of the school year: on January 26, 30, and 31. The third graders had been working for several days on ideas related to even and odd numbers,

[3] In describing and discussing this episode I use with permission parts of Stylianides (2007c) and Stylianides and Ball (2008) (license numbers 3487830512791 and 3454130366327, respectively).

and they had formulated some conjectures with Ball's help. One of those conjectures was called "Betsy's conjecture" after the student who proposed it:

1. *Betsy's conjecture*: "An odd number plus an odd number equals an even number."

On January 26 the class worked on this conjecture. The children had provided examples of the conjecture before Betsy suggested revising the conjecture to be about the sum of the *same* odd number:

2. *Betsy*: Okay, now there's one thing on my conjecture that I have to revise. [...] I am trying to show that this has to mean the *same* odd number, not two different odd numbers.
3. *Ball*: What do other people think? Does it have to be the same odd number, or could you use two different odd numbers? Look in your notebooks and see what you tried. Our examples on the board are all the same odd number, right? Seven plus seven is the same one. Thirty-five plus thirty-five is the same one. Do you agree with her? Should it be revised so it says it has to be the same odd number?

Tembe objected to the idea of revising the conjecture and he offered the example $7+9=16$. But in showing this example he became confused, thinking that he had refuted the conjecture. Yet others realized that Tembe's example actually fitted the conjecture and thus did not justify its revision.

4. *Betsy*: Okay, yeah, you can use two different...
5. *Ball*: He said seven plus nine is sixteen. Does that fit the conjecture or not?
6. *Students*: Yeah, yes, it does!
7. *Ball*: Who can say why it does? How about... how about Sheena? Why does that fit?
8. *Sheena*: Because if you have two different numbers like seven and nine, it still has to be an even, um it still has to be, it's still two odd numbers and it says an odd number plus an odd number equals an even number, and sixteen is an even number. So it's the same thing.

The class dropped the idea of revising the conjecture, and Ball invited further comments from the students about the conjecture:

9. *Ball*: Other comments about the conjecture? [...] Did anybody come up with an example that didn't work, for example? Did anybody come up with an odd number plus an odd number that *didn't* equal an even number? Jeannie, you found one?
10. *Jeannie*: Me and Sheena were working together, but we didn't find one that didn't work. We were trying to *prove* that, um, Betsy's conjecture, um, that you *can't* prove that Betsy's conjecture always works [murmurs from other children].
11. *Ball*: Go on, Jeannie. Say more about why you think that.
12. *Jeannie*: Because um there's um like numbers go on and on forever and that means odd numbers and even numbers, um, go on forever and, um, so you couldn't prove that *all* of them work.

13. *Ball*: What are people's reactions to what Jeannie and Sheena thought? [pause] They said they didn't find one that didn't work, but they don't think we can prove it *always* works because numbers go on forever and ever.
14. *Ofala*: I think it can always work because I um tried almost, um, [she counted the examples she had in her notebook] eighteen of them, and I also tried a Sean number [a class-invented term used to describe an even number with an odd number of groups of two] so I think, I think it can always work.

Mei also disagreed with Jeannie, saying that Jeannie had accepted other conjectures formulated by the class without, however, insisting that these conjectures were shown to be true for all possible cases:

15. *Mei*: [...] Because with those conjectures [she motioned to several previously discussed and widely agreed upon conjectures that were posted above the board] we haven't even tried them with all the numbers that there is, so why do you [Jeannie] say *those* work? Well, we haven't even tried all those numbers that there ever could *be*. [...] So you can't really say that *those* are true if you're saying that you want to try every number there ever was.
16. *Ball*: Jeannie?
17. *Jeannie*: I never *said* that they were true all the time...
18. *Mei*: Then why do, then why didn't you disagree when like Lucy and everybody agreed with those conjectures?
19. *Jeannie*: Because I wasn't *thinking* about it.
20. *Sheena*: She just thought about it. That's why we were trying to think about it *today* [...]

The discussion continued for a few minutes, with some children explaining that they had tested many examples and this was quite convincing. In preparing for the next lesson on January 30, Ball wrote in her journal (January 30, p. 195):

21. How does a mathematician try to prove this [Betsy's conjecture]? By showing that there's something *logically necessary* about odd numbers that would make them, when combined, always equal an even number. Would/Could third graders accept/understand that because all odd numbers are of the form $2n + 1$ that the two "loose" ones would always combine to make a group of two? I'd like to experiment with whether they *can* understand this—I wonder what I could do to push them on this. [emphasis in original]

Ball began the lesson by reviewing Jeannie and Sheena's idea that the class could not prove Betsy's conjecture because numbers (odd and even) are infinite. She also engaged the students in clarifying Jeannie and Sheena's idea by having them describe the idea in their own words.

Several students thought Betsy's conjecture was always true, because they had checked many examples that verified it, but some others began to realize the complexity in trying to prove the conjecture for *all* pairs of odd numbers. This second group of students started to see that, no matter how many examples one found that verified the conjecture, there still might be a pair of odd numbers for which the conjecture failed. At some point Ball said:

22. *Ball*: This is an important thing that Jeannie and Sheena brought up. When mathematicians make an idea like this [Betsy's conjecture], they want to be able to prove that it *always* works. They don't like the idea of having ideas saying, "Well, we've tried it a little bit, we can't be sure." And so they work pretty hard to see if they can prove it. But we know that mathematicians don't write down every last number, right?
23. *Students*: Yeah!
24. *Ball*: Why couldn't they write down every last number? [pause] Nathan?
25. *Nathan*: Because they don't know every last number.
26. *Ball*: Why don't they know them all? I thought they were mathematicians! [Students laughed.]

Nathan explained that mathematicians would have to go from tens to hundreds, to thousands, to millions, and so forth, until they "[did] not know what [came] next or how to pronounce it right." Ball asked for another reason why mathematicians would not be able to write down every number. Keith said that numbers "go on and on," so one could never finish recording all numbers. Ball then said:

27. *Ball*: So mathematicians need something they can do. I wanted to show you something that a mathematician would try to do when trying to prove something like this. Okay? Let's see if we can work with something that a mathematician would try to do. Mathematicians when they try to prove something like this would try to say: "If there is something about odd numbers and something about even numbers that I can understand that would make this *always, always* be true, without having to try every single number." And they might try to do something like this. I was thinking about a picture that Ofala made a while ago. Ofala made a picture of [pause] when she was trying to give us the definition for an odd number, she made a picture [pause] I can't remember what numbers she tried, but let's say she tried 7. [She drew seven vertical lines on the board.] Do you remember what Ofala did to prove that it was odd?

Several students volunteered to reproduce Ofala's explanation. This brought to the fore the classroom community's (primary) definition of odd numbers:

28. *Definition of odd numbers*: "Numbers that if you group them by twos there is one left over." (The implied domain of reference is the set of natural numbers.)

The students also illustrated how this definition would apply for the number 7 (Figure 7.3).

Ball also helped the students remind themselves of their (primary) definition of even numbers:

29. *Definition of even numbers*: "Numbers that if you group them by twos there is nothing left over." (The implied domain of reference is the set of natural numbers.)

Ofala drew a picture on the board, like the one in Figure 7.4, to show how one could use these definitions to illustrate the conjecture in the particular case $7 + 9 = 16$. Ball used Ofala's picture to explain that the two "loose" ones from each odd number combined to make another group of two. She also said that a mathematician would think hard how to use these properties of odd and even numbers to convince other people that this would happen with *any* pair of odd numbers. She then asked the class to use pictures to show other instances of the conjecture.

At some point Lucy said she had an "exception" to the conjecture and presented the vertical addition in Figure 7.5(a). Betsy observed that Lucy did not line up the two numbers correctly, which resulted in adding hundreds with millions. Betsy corrected Lucy's addition as shown in Figure 7.5(b), thus showing that the sum was an even number. Lucy was convinced by Betsy's argument and recognized that her example was not really an exception to the conjecture.

In deciding about the parity of 1,000,108, Betsy drew on another definition of even numbers that was used by some students in the class:

30. *Another definition of even numbers*: "Numbers that end up in 0, 2, 4, 6, or 8." (The implied domain of reference is the set of natural numbers.)

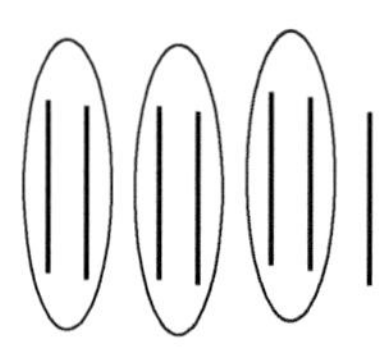

Figure 7.3 An illustration that number 7 is an odd number.

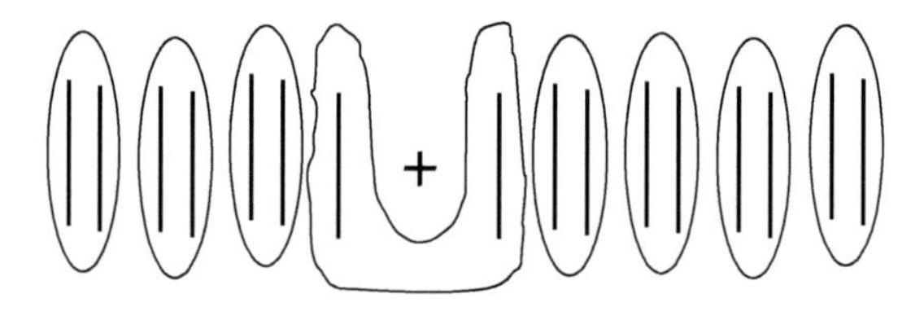

Figure 7.4 An illustration of the conjecture "odd plus odd equals even" in the particular case $7 + 9 = 16$.

(a)	(b)
101	101
+ 1000007	+ 1000007
2010007	1000108

Figure 7.5 Lucy's "exception" to Betsy's conjecture (a) and the corrected calculation (b).

Ball started the lesson on January 31 by eliciting reports from different groups of students on their work the previous lesson. Tembe went up to the board first and illustrated Betsy's conjecture in the particular case $7+7=14$. After that Betsy volunteered to show a "proof," as she called her argument:

31. *Betsy*: [She went up to the board and said:] What we figured out how it's always true is that we would have seven dots, or lines plus seven lines [she drew 14 lines on the board as shown in Figure 7.6a]... and then [she counted the lines]... then we said that we had to circle them by twos [she started circling pairs of lines]... and also we said that . . . just a second [she finished circling all pairs of lines as shown in Figure 7.6b].

32. *Betsy (continued)*: [From this point onward Betsy faced the class, away from the board, as she talked.] That if you added another even one to an odd number, or another one to an odd number, then it would equal an even number, 'cause all odd numbers if you circle them, what we found out, all odd numbers if you circle them by twos, there's one left over, so if you... plus one, um, or if you plus another odd number, then the two ones left over will group together, and it will make an even number.

Betsy's argument generated discussion in the class about whether the argument actually *proved* that the sum of any two odd numbers would be even. After a few questions from students Betsy said:

33. *Betsy*:. . . . If you added two odd numbers together you can add the ones left over and it would always equal an even number.

Ball then summarized the discussion up to that point and asked for other reactions from students:

34. *Ball*: Tembe gave one example and Betsy tried to show that this would always be true. . . . Do people think that [Betsy's argument] *does* prove that an odd plus an odd would always be even?
35. *Riba*: No. [Mei nodded her head in agreement with what Riba just said.]

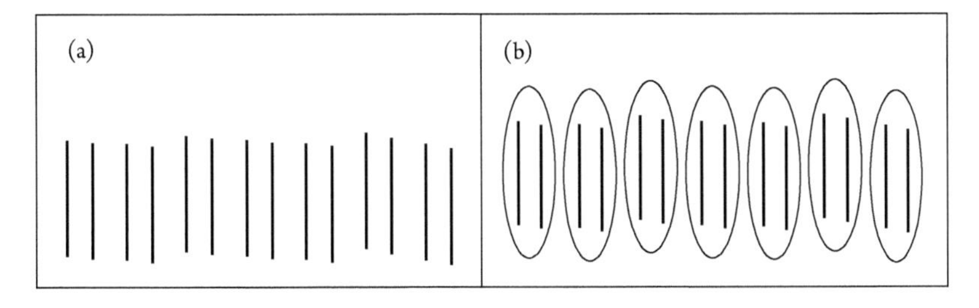

Figure 7.6 Betsy's writing on the board at the beginning of the presentation of her proof.

36. *Ball*: Why not, Mei?
37. *Mei*: I don't think so, because you don't know about, like, in the thousands, and you don't know the numbers, like, you don't even know how you *pronounce* it, or how you *say* it.
38. *Ball*: What do other people think? Does the thing that Betsy did, does that show that any time you have an odd number plus an odd number, you get an even number? Mei thinks it doesn't because you wouldn't know about big, very big numbers. [...]
39. *Jeannie*: I don't think so, because she [Betsy] didn't say it had to *be* those two numbers [she referred to Tembe's example and Betsy's illustration of it presented in Figure 7.6], it could be *any* two odd numbers because, um, there's always one left [over].
40. *Mei*: I know, but that is not, like, telling.... I'm trying to say that that's only one example. You can't really say that it will work for every odd number, even—I don't think it would work for numbers that we can't say or figure out what they are.
41. *Ball*: What do other people think about this argument that Jeannie and Mei are having? Mei's saying that she is still not convinced that this could apply for all numbers, big, big numbers, and Jeannie is saying that it *would* always be true because there'd always be one left over from each odd number. Riba, what do you think about that?
42. *Riba*: How does, um, Jeannie know that it would always work? She never tried all the numbers.
43. *Betsy*: Mathematicians can't even do that. You would die before you counted every number.
44. *Riba*: I know, that's why I mean, that's what I mean—you don't know if it *always* works.
45. *Ball*: Anyone else have a comment? Sheena?
46. *Sheena*: I agree sort of with Riba, but you don't have to try *all* the numbers, because you *would* die before you tried all the numbers because there are some numbers you can't even pronounce, and some numbers you don't even know that they're *there*. But, still, I still think it's true because, um, we've tried like enough of, of examples already. It proves it's true.

The discussion continued for a while and was followed by more work on the conjecture in small groups of students. When the class resumed, Ball called on Mark to report on his work with Nathan.

47. *Mark*: Well, first, me and Nathan, we were just getting answers and we weren't thinking about proof. And we were still getting answers and we were thinking about, we were trying to prove, and Betsy came and she had proved it, and then we all agree that it would work.

The lesson continued with the students working on a different conjecture.

Discussion of the Episode

The proving task in this episode concerned Betsy's conjecture (line 1), which was about the parity of the sum of any pair of odd numbers and thus involved an infinite number of cases. Because the students did not know whether the conjecture was true or false, issues of both *refutation* and *justification* arose naturally in their proving activity.

The issue of the possible refutation of the conjecture came up twice during the episode. The first time was when Betsy suggested revising the conjecture to be only about reflexive pairs of odd numbers (line 2). Betsy seemed to think that the sum of two different odd numbers would not (necessarily) be even. Ball invited comments from the class on this issue (line 3) and Tembe offered an example of two different odd numbers that added to an even number. After some prompting from Ball (lines 5 and 7), Sheena explained that Tembe's example supported no revision of the conjecture (line 8). The issue of the possible refutation of the conjecture came up for the second time when Lucy said she had an "exception" to the conjecture. As it turned out, however, this was not a valid counterexample: Lucy had made a place value error, which produced a computational error (Figure 7.5a).

The issue of whether and how the conjecture could be justified played out prominently in this episode, placing varied and rather heavy demands on the teacher as she was trying to support students' work. The issue was brought up for the first time, and in a rather dramatic way, by Jeannie who, in reporting her work with Sheena, said that the class could *not* "prove that Betsy's conjecture always work[ed]," because "numbers go on and on forever" and one could not "prove that *all* of them work[ed]" (lines 10 and 12). Jeannie and Sheena's idea seemed to be the following: A proof of Betsy's conjecture would have to consider all cases involved in it, and this can be done only by checking every single case separately; the conjecture, however, involves an infinite number of cases, and so it is impossible for one to prove it. Jeannie and Sheena's idea possibly embodied the realization that *empirical arguments* do not meet the standard of proof: checking some but not all possible cases covered by the conjecture would not meet these students' expectations of a proof to show that *all* cases worked.

Jeannie and Sheena's idea had a notable mathematical and pedagogical value, and this is probably why Ball gave it such a prominent place in the work of the class. Specifically, she organized a whole class discussion around the idea (lines 13–20); she reflected on its mathematical and pedagogical implications when planning for the next lesson (line 21); and she began the new lesson by reviewing the idea, asking students to describe it in their own words, and using it as a starting point to introduce students to mathematicians' way of proving statements over infinite sets (from line 22 onward). From a mathematical standpoint, the idea implied the need for students to consider *all* possible cases in proving mathematical statements over infinite sets, thus also implying the inadequacy of empirical arguments as proofs of mathematical generalizations. From a pedagogical standpoint, the idea highlighted the importance of helping students appreciate the notion of an *infinite* number of cases, and it implied the need to equip students with mathematical tools that would allow them to justify mathematical statements over infinite sets without resorting to the impossible method of checking every single case separately.

Following Ball's invitation for reactions to Jeannie and Sheena's idea when this was first expressed (line 13), the response of the class, as reflected in the contributions of Ofala and Mei (lines 14, 15, and 18), was that of objection. Ofala's objection was based on the view, grounded implicitly on an empirical argument, that the conjecture was true (line 14). Mei's objection was based on the view that Jeannie and Sheena's idea contradicted what Mei considered to be established practices in classroom work regarding the acceptance and use of conjectures (lines 15 and 18). According to Mei, up until that point the class had accepted and used conjectures without requiring that they were checked for all possible cases, and so she saw a contradiction in Jeannie and Sheena now questioning practices that they themselves had also approved, or tolerated at the very least, for so long. This is an interesting argument: Although it was formulated on social rather than mathematical grounds, thus fitting more under the framework of argumentation (e.g., Boero et al., 1996; Krummheuer, 1995; Mariotti, 2006), it was essentially an argument by *reductio ad absurdum*. The argument was implicitly based on the assumption that questioning a classroom practice cannot be an afterthought, but rather it should be raised at the point when the practice is being established.

Ball started the lesson on January 30 by inviting students to describe, rather than react to, Jeannie and Sheena's idea, thus creating a constructive context for the class to explore the idea and its implications. Ball tried to help the students understand that one could not possibly check *all* pairs of odd numbers due to their infinite number (see lines 22–26 and also Ball's exchanges with Nathan and Keith). Students described and understood (to some extent) the notion of an infinite set such as that of natural numbers with respect to its unboundedness; in Jeannie's words "numbers go on and on forever" (line 12). Furthermore, and being aware of the important role of definitions of odd and even numbers in addressing the challenge of how to construct a general argument for Betsy's conjecture (line 21), Ball brought the issue of definitions to the center of students' work (line 27 and after). She explained to the students that mathematicians would use definitions in trying to understand what makes a conjecture like Betsy's to "*always, always* be true, without having to try every single number" (line 27). She also made sure that the students reminded themselves of their definitions of odd and even numbers (lines 28 and 29), and she posed to them the challenge to think how they could use these definitions to prove Betsy's conjecture.

Ball's scaffolding yielded results on January 31, when Betsy presented the following argument for the conjecture. This is an excerpt from a more extended contribution that Betsy made in front of the class (lines 31–32). The first half of Betsy's contribution (line 31) was a pictorial illustration (Figure 7.6) of a particular example that Tembe had written earlier on the board and might reasonably be considered to be separate from Betsy's argument in line 32:

> [A]ll odd numbers if you circle them, what we found out, all odd numbers if you circle them by twos, there's one left over, so if you . . . plus one, um, or if you plus another odd number, then the two ones left over will group together, and it will make an even number.

According to the definition of proof described in Chapter 2, this argument qualified as a proof for the conjecture in the particular classroom context because:

(1) It used (explicitly or implicitly) true statements that were readily accepted by the classroom community, notably, the community's definitions of odd and even numbers.

(2) It employed a valid mode of argumentation that appeared to be within the reach of the classroom community and was associated with the use of definitions to deduce logically the statement of the conjecture.

(3) It was represented using verbal language in a way that was appropriate and the students in the class could understand.

With regard to point (1), which concerns the classroom community's set of accepted statements, the argument referred explicitly to the community's definition of odd numbers (line 28) but only implicitly to the definition of even numbers (line 29). The use of the definition of even numbers would have been made explicit if Betsy had finished her argument by saying something like, " . . . because you can group the resulting number by twos with nothing left over."

With regard to point (3), which concerns the mode of representation of Betsy's argument, one might observe that Betsy's verbal language essentially described in words the symbolic language that is used in the standard algebraic proof that students are usually taught or expected to produce at secondary school. In Table 7.1 I present the correspondences between these two arguments using different modes of representation: verbal language and algebraic notation. The sound correspondences between the two arguments, both of which qualify as proofs, illustrate the "elastic" nature of proof discussed in Chapter 2, according to which what counts as a proof can grow with students' knowledge while preserving some core mathematical qualities (notably, valid modes of argumentation).

Table 7.1 Correspondences between Betsy's argument and the standard algebraic proof.[a]

Betsy's argument	Standard algebraic proof
[A]ll odd numbers if you circle them, what we found out, all odd numbers if you circle them by twos, there's one left over	All odd numbers are of the form $2n+1$, where n is a whole number
[S]o if you . . . plus one, um, or if you plus another odd number, then the two ones left over will group together,	So if you add two odd numbers, you get $(2k+1)+(2m+1)=(2k+2m)+(1+1)$
and it will make an even number	and it will make an even number
[*Implicit in the last step is the reason: "because you can group the resulting number by twos with nothing left over." This explanation uses the definition of even numbers as the numbers that when grouped by twos there is nothing left over.*]	[*Implicit in the last step is the reason: "because the expression can be written in the form $2(k+m+1)$." This explanation uses the definition of even numbers as the numbers of the form $2n$, where n is a natural number.*]

[a]Both arguments are formulated over the set of natural numbers.

Further, with regard to point (3), while Betsy's argument was represented in a way that the students in the class could understand, the discussion that followed Betsy's presentation (lines 34–46) suggests that some students in the class might have misunderstood what the presented argument actually was. Specifically, Mei seemed to have associated Betsy's argument with the particular case that Betsy had illustrated first on the board (line 31 and Figure 7.6) rather than with the general argument she presented afterward (line 32). This interpretation of Mei's response to Betsy's argument follows from Mei's explanation of her objections to the argument (lines 37 and 40), especially when she said: "I'm trying to say that that's only one example. You can't really say that it will work for every odd number" (line 40). This interpretation is corroborated by how Jeannie understood and responded to Mei's objections, especially Jeannie's clarification to Mei that Betsy's argument was about *any* two odd numbers: "[Betsy] didn't say it had to *be* those two numbers [on the board], it could be *any* two odd numbers because, um, there's always one left [over]" (line 39).

Yet it is also possible that Mei's objections to Betsy's argument related to a difficulty in understanding the mode of argumentation used in Betsy's argument, especially how the idea of "grouping by twos" could apply to numbers one "[didn't] even know how [to] *pronounce*" (line 37) or "figure out what they [were]" (line 40). Riba also seemed to have difficulty in understanding or accepting the mode of argumentation used in Betsy's argument: Riba could not see how the argument covered all possible cases given that it did not actually check every single pair of odd numbers (lines 42 and 44). Difficulty in understanding or accepting the particular mode of argumentation is reflected further in Sheena's contribution (line 46), though unlike Mei and Riba, Sheena did not claim or suggest that Betsy's argument was invalid. Rather, she proposed an empirical argument, considering that the conjecture was "true because, um, [the class had] tried like enough of, of examples already" (line 46).

Interestingly, Sheena's empirical argument appears to contradict my earlier analysis of the idea that she and Jeannie had expressed at the beginning of the episode (lines 10 and 12), namely, my inference that their idea possibly embodied the realization that empirical arguments are not proofs. This apparent contradiction is puzzling, but it may nevertheless be explained in three ways. First, while the idea emerged from the work of both Sheena and Jeannie, it is possible it belonged more to Jeannie, who was also the person who presented and defended the idea to the rest of the class. Indeed, Jeannie's overall activity during the episode, including the way she supported Betsy's argument (line 39), was consistent with the idea she had presented and defended at the beginning of the episode. Second, Sheena's possible realization that "empirical arguments are not proofs" at the beginning of the episode might have been fragile, thus giving way, by the end of the episode, to the common student misconception that "empirical arguments can be proofs." Third, Sheena's use of the empirical mode of argumentation at the end of the episode might have been underpinned by a difficulty in understanding the mode of argumentation used in Betsy's argument or in thinking independently of a viable alternative to the empirical mode of argumentation.

Students' difficulties with the mode of argumentation used in Betsy's argument might appear to suggest that this mode was beyond the classroom community's reach at the given time. I would challenge that on the basis of evidence from this episode: One student in the class—Betsy—used the particular mode, while at least three other students—Jeannie,

Mark, and Nathan—seemed to understand or appreciate it. In particular, Mark and Nathan admitted that Betsy's argument caused a shift in their approach to proving: from "just getting answers" and not "thinking about proof"; to "trying to prove"; to recognizing that "[Betsy] had proved [the conjecture]" and "agree[ing] that it would work" (line 47). Further evidence that the particular mode of argumentation is, or can be, within young children's reach is found in reports from other supportive elementary classrooms where students generated similar arguments for the same conjecture (e.g., Carpenter et al., 2003[4]), as well as from psychological research on the cognitive development of young children's ability for deductive reasoning (reviewed in Stylianides & Stylianides, 2008b). Of course, active involvement of the teacher is required to enable young students to understand, appreciate, or use the particular mode of argumentation (Hanna & Jahnke, 1996; Stylianides, 2007b).

In managing the whole class discussion around Betsy's argument (lines 34–47), Ball helped her students attend to and reflect on Betsy's argument. Specifically, she distinguished Betsy's argument from the specific example offered by Tembe when she first invited students to comment on whether Betsy's argument qualified as a proof (line 34), and she kept the discussion focused on this important issue while also re-voicing key points in students' contributions (lines 38 and 41). The discussion turned out to be a debate between two groups of students: on the one side there were students such as Mei and Riba who objected to Betsy's argument being considered a proof; on the other side there were students such as Betsy and Jeannie who tried to rebut the objections and further clarify aspects of the argument that seemed to have been missed or misunderstood by the opposing side. Although the work of the class on the conjecture finished with a strong statement from Mark and Nathan endorsing Betsy's argument as a proof for the conjecture (line 47), it is likely that several students were still in a position of disbelief or objection. This raises the issue of what the role of the teacher is, or might be, in facilitating the social process of an argument that meets the standard of proof to be accepted as such by the classroom community. I return to this issue in the general discussion at the end of the chapter.

To conclude, the proving task in this episode involved an infinite number of cases and helped engage students in proving activity whose primary purpose at any given moment—to justify or to refute Betsy's conjecture—placed different demands on the teacher as she managed students' work. When students tried to refute, or thought they had refuted, the conjecture, which was impossible as the conjecture was actually true, things were relatively unproblematic for the teacher: She helped the students interpret more accurately their examples, or correct the calculation errors in their examples, thus leading them to the realization that they offered no disconfirming evidence for the conjecture. The issue of whether and how the conjecture could be justified created a lot of challenges for the class. While some students began to realize that justifying the conjecture required a method to consider all (infinitely many) cases and they raised the question whether such a method actually existed, some other students considered that empirical arguments offered adequate validation methods. The teacher made persistent efforts to help the class appreciate the

[4] Carpenter et al. (2003, p. 90) reported the following argument of a student called Jamie: "Well, for any odd number, you have groups of two with one left over. So the two odd numbers have two leftovers, and you can put them together, so there are no leftovers that are not paired up. So it's even."

importance of considering all possible cases involved in a conjecture believed to be true, and she introduced the class to the mode of argumentation associated with the use of definitions to deduce logically a statement over an infinite set. This episode illustrated that, although accepting or using this mode of argumentation can be challenging for elementary students, at least when they are first introduced to it, the mode can nevertheless be accessible to them given a supportive classroom environment and scaffolding from the teacher.

General Discussion

Episodes G and H offered complementary views on what it might mean or look like when elementary teachers engage their students with proving tasks that involve an infinite number of cases. In this final section I discuss some general issues about the relationship between proving tasks and proving activity, in the particular context of proving tasks that involve infinitely many cases, and about the role of the teacher while implementing this kind of task in the classroom.

The Relationship between Proving Tasks and Proving Activity

The definition of proof that I discussed in Chapter 2 breaks a mathematical argument down into three components: the set of accepted statements, the modes of argumentation, and the modes of argument representation. The second of these components played a key role in the proving activity generated by the proving tasks in these two episodes; the truth values of the conjectures that were explored in each episode, coupled with students' initial unawareness of whether the conjectures were true or false, helped bring to the fore different modes of argumentation.

In Episode G the exploration of several false conjectures raised the issue about what was involved in refuting a false statement that was formulated over an infinite set, thus bringing to the fore the mode of argumentation associated with refutation by means of a counterexample. While the students in the episode seemed to recognize that counterexamples offered disconfirming evidence for a false conjecture, thus necessitating its revision, they had difficulty in recognizing the power of a single counterexample to refute a false mathematical generalization. This observation is not surprising: Several studies reported that students or adults tend to consider a single counterexample to be insufficient to establish the falsity of a mathematical generalization (e.g., Balacheff, 1988b; Mason & Klymchuk, 2009; Simon & Blume, 1996; Zaslavsky & Ron, 1998), while other studies reported a broader set of difficulties that even mathematics majors tend to have with proof by counterexample (e.g., Ko & Knuth, 2009, 2013; Leung & Lew, 2013). The issue of justifying true mathematical generalizations also emerged in Episode G (once the original conjectures had been revised to rule out exceptions), but this issue played out more prominently in Episode H, to which I turn next.

In Episode H the conjecture explored by the class was true and, despite a few attempts by students to refute the conjecture, the bulk of the activity was about whether and how the conjecture (formulated over an infinite set) could be justified. This brought to the fore two different modes of argumentation: first, the invalid mode associated with empirical arguments, whereby the conjecture could be accepted as true based on the confirming (albeit inclusive) evidence offered by the examination of a few cases in its domain; second, the valid mode associated with the use of definitions to deduce logically the statement of the conjecture.

Students' use of the first mode of argumentation is unsurprising, given that even advanced secondary students and adults were found to consider empirical arguments as proofs of mathematical generalizations (e.g., Harel & Sowder, 2007; Küchemann & Hoyles, 2001–2003; Morris, 2007). Also, the way in which some students in Episode H used the empirical mode of argumentation offers supportive evidence for a hypothesis I made in Chapter 6. Specifically, I hypothesized in that chapter that students' empirical arguments for true statements over *infinite* sets might be based, at last partly, on their attempts to use a mode of argumentation that is relevant to the justification of true statements over *finite* sets, namely, considering separately all possible cases involved in the set. The impossibility of using this mode of argumentation to justify true statements over infinite sets, combined with students' lack of knowledge or understanding of a viable alternative mode of argumentation, might have led some of them, such as Sheena (line 46), to resort to the empirical mode of argumentation. Indeed, given the pressure that the "didactical contract" (Brousseau, 1997) usually puts on students to produce an argument for any proving task that is set before them by the teacher, it is not unreasonable for students to resort to empirical arguments when they are unable to produce better arguments, even in those cases when they are fully aware of the limitations of such arguments (Stylianides & Stylianides, 2009a).

The second mode of argumentation exemplified in Episode H was associated with the use of definitions to deduce logically statements over infinite sets. This was valid and a viable alternative to the empirical mode of argumentation. It was used in Betsy's argument and was appreciated by a few students in the class, though there were also several other students who were challenged by it. Students' difficulties with this mode of argumentation are unsurprising, not only because the class had just been introduced to it (see line 27 and onward), but also in light of research findings concerning older students' difficulties with modes of argumentation used in or required by proofs of generalizations over infinite sets (a recent review of some of these findings is found in Stylianides et al., 2016b). A mode of argumentation similar to that in Betsy's argument was used also in Episode G during the presentation of Jack's argument, but, as I have already discussed, the teacher played a key role in that presentation.

Overall, the proving activity in Episode G illustrated what it might mean or look like when elementary students engage in exploration and refutation/revision of false generalizations over infinite sets (see also Reid, 2002), including student challenges in understanding what is involved in establishing that a generalization is false. Episode H illustrated what it might mean or look like when elementary students come up against, and grapple with, the challenges of justifying that a generalization over an infinite set is true. In addition to exemplifying the challenges encountered by students, both episodes highlighted the important role of the teacher in supporting students' proving activity and in facilitating the use of, or bringing within students' reach, the specific modes of argumentation required for a proof. I shall now elaborate on the teacher's role.

The Role of the Teacher

These episodes illuminate several aspects of the role of a teacher who implements a proving task that involves refutation or justification of a mathematical generalization over an infinite set. I discuss three of these aspects.

The first aspect is that, when planning the implementation of one such task in an elementary classroom, the teacher needs to size up students' capacity to use or understand the specific modes of argumentation required to refute or justify the mathematical generalization at hand. These two episodes, as well as the findings of relevant research referenced earlier, allow little doubt that an elementary teacher who implements a proving task involving a generalization over an infinite set will have to deal with major student difficulties, such as difficulty in recognizing the adequacy of a single counterexample to refute a false generalization (Episode G) or the insufficiency of a few confirming examples to justify a true generalization (Episode H). Anticipating difficulties that students can face during the implementation of a cognitively demanding task is of course important (e.g., Stein et al., 2008), and, as a field, we can predict quite securely students' difficulties with modes of argumentation when they engage with proving tasks that involve mathematical generalizations over infinite sets. However, we know little about effective ways to help students overcome these difficulties (Stylianides et al., 2016b). To make matters worse, the few research findings currently available on this topic are mainly from studies with secondary or university students and thus cannot readily inform teaching practice at elementary school. For example, a research-based instructional intervention that we found can help university students (pre-service elementary teachers) and secondary students to begin to recognize the limitations of the empirical mode of argumentation (Stylianides, 2009a; Stylianides & Stylianides, 2009b, 2014b) cannot be easily adapted for use with elementary school students, not least because some of the proving tasks in the intervention are beyond young children's reach.

Indeed, more research is needed to develop effective teaching practices in elementary classrooms to help students appreciate and use modes of argumentation for refuting or justifying generalizations over infinite sets. But, more fundamentally, research is also needed to understand what makes this appreciation or use difficult for elementary students in particular. For example, it is unclear why the change made by Howard to the adverbs of quantification in the statement of a conjecture in Episode G, from "always" to "sometimes," appeared to resolve students' difficulties in recognizing the correct truth value of the conjecture. Research on students' understanding, typically misuse, of quantifiers has focused on older students (e.g., Arsac & Durand-Guerrier, 2005; Epp, 2009) and offers little insight into how elementary students' understanding of adverbs of quantification such as the above might interfere with their understanding of the statements of mathematical generalizations over infinite sets.

The second aspect of the role of the teacher relates to the first. Although there is a thin research basis about how elementary teachers can prepare themselves for the challenges that are likely to emerge when they implement proving tasks involving infinitely many cases, the teachers still have to make their own decisions about how to intervene in students' proving activity so as to help them access the accepted modes of argumentation. I view this interventionist aspect of the teacher's role as supportive, not dismissive, of students' constructive efforts. Indeed, "the suggestion that students can be left to their own devices to construct the mathematical ways of knowing compatible with those of a wider society is a contradiction in terms" (Cobb, Yackel, & Wood, 1992, p. 27). This is particularly relevant to the area of proving where a passive role for the teacher would mean that "students are denied access to available methods of proving," for "[i]t would seem unrealistic to expect students to rediscover sophisticated mathematical methods or even the accepted modes of argumentation" (Hanna & Jahnke, 1996, p. 887).

The teachers in these two episodes intervened in some decisive ways in students' proving activity, acting as representatives of the discipline of mathematics in their classrooms and as the people charged with the responsibility of connecting their students with conventional mathematical knowledge (cf. Ball, 1993; Lampert, 1992; Stylianides, 2007b; Yackel & Cobb, 1996). For example, in Episode G Howard took a leading role in presenting Jack's generic argument to the rest of the class and helped the students develop it into a more general proof using algebraic notation. In Episode H Ball introduced the students to the mathematicians' way of using definitions to prove general statements over infinite sets. It is important to note, however, that while implementing these actions the two teachers built on or extended specific student contributions, thus contingently responding to emerging student needs for them to access conventional knowledge. If the teachers had not responded to those needs, which of course were more implicit than explicit in students' contributions, important learning opportunities for students would have been missed. Take, for example, Jeannie and Sheena's idea in Episode H that a proof of Betsy's conjecture might not be possible due to the infinite number of cases involved in it and the impossibility of checking every single case separately. This idea essentially reflected a need for students to learn, and for the teacher to teach them, an alternative mode of argumentation that would allow them to justify true statements over infinite sets without resorting to the impossible method of separate checks. If the students had been left to their own devices it would be unrealistic to expect that they would have discovered such a mode of argumentation. The teacher's actions in the episode to induct the students into such a mode of argumentation fit in well with the definition of *scaffolding* as "the process that enables a child or novice to solve a problem, carry out a task, or achieve a goal which would be beyond his [or her] unassisted efforts" (Wood, Bruner, & Ross, 1976, p. 90). The teacher's actions seemed to have enabled the development of Bernadette's proof later in the episode, thus also satisfying another common feature of scaffolding: "to bring the learner closer to a state of competence which will enable them eventually to complete such a task [i.e., one which was originally beyond their unassisted efforts] on their own" (Maybin, Mercer, & Stierer, 1992, p. 188).

The third aspect of the role of the teacher relates to facilitating the social process of an argument that has met the standard of proof to actually be accepted as a proof by the classroom community. In Episode G, as in episodes discussed in previous chapters, several arguments that met the standard of proof received no objections from the class, and so this created no pressing need for the teacher to raise or pursue the issue of social acceptance of these arguments as proofs. In Episode H, however, an argument that met the standard of proof received a reaction of disbelief or objection from students in the class, with several of them likely staying unconvinced by the argument until the end of the lesson. What might be the role of the teacher in this kind of situation in the classroom?

In thinking about this issue it may be useful to see first what happens in the mathematical community. It has been remarked that in the mathematical community "[a] proof becomes a proof after the social act of 'accepting it as a proof'" (Manin, 1977, in Hanna, 1983, p. 71) and that "proof is *convincing argument, as judged by qualified judges* [peer mathematicians]" (Hersh, 1993, p. 389; emphasis in original). Clearly, conviction in the mathematical community is not an arbitrary process; rather, it rests on socially accepted rules of discourse that ensure the quality of arguments that are accepted as proofs (de Millo, Lipton, & Perlis, 1979/1998). In a classroom community, however, the process of acceptance of a proof could not be based as heavily on conviction as happens in the mathematical community, for that would presuppose prior acceptance among classroom participants of rules of discourse that would help winnow out the arbitrariness of conviction (Stylianides, 2007c). The teacher's role might involve, then, facilitating the negotiation of rules of discourse among classroom participants and trying to ensure that, over time, the results of this negotiation approximate those in the mathematical community (Balacheff, 1990, 1991; Lampert, 1990; Simon & Blume, 1996; Stylianides, 2007c; Yackel & Cobb, 1996). Research is needed to better understand this role, including the many challenges that the role entails (see, e.g., Balacheff, 1991).

8

Conclusion

In this chapter I draw some general conclusions in light of my discussions in the previous chapters, notably Chapters 4–7. Specifically, I look across these chapters and try to distill some key insights in relation to the two core issues considered in this book: (1) the relationship between different kinds of proving tasks and corresponding proving activity, and (2) the role of the teacher as a mediator of this relationship. In doing so, I also identify a few major directions for future research, which supplement those identified in previous chapters. Finally, I revisit the book's general aim (described in Chapter 1) and discuss implications of the work reported herein for how the place of proving in elementary students' mathematical work might be elevated.

The Relationship between Proving Tasks and Proving Activity

In setting up the investigation of the relationship between proving tasks and proving activity in Chapter 3, I acknowledged that the relationship is complex and defies general rules. At the same time, however, I proposed that the varying characteristics of the kinds of proving tasks that I identified could support qualitatively different proving activities during their implementation in the elementary classroom. In this section I revisit and discuss that proposal.

To support my discussion, I summarize in Table 8.1 the main arguments that were offered or implied during the proving activities described in the eight episodes in Chapters 4–7. Each argument in the table is associated with a particular mode of argumentation (valid or invalid) and is a distilled version of one or more related arguments that students generated, independently or with support from the teacher, as they worked on different kinds of proving tasks.

One could say that the arguments in the table could have been derived on the basis of a mathematical analysis of the affordances of the different kinds of proving tasks, in tandem with an analysis of relevant findings from mathematics education research on different kinds of student arguments. While I appreciate the merit of these two analyses, and

Table 8.1 The main arguments—valid and invalid (in italics)—offered or implied during the proving activities described in the eight episodes (Chapters 4–7), in the context of different kinds of proving tasks.

Number of cases involved in a proving task	Purpose of a proving task	
	Justification of a statement	**Refutation of a statement**
A single case	(a) To justify an answer to a problem involving a single calculation, use established procedures (e.g., the steps of a calculation algorithm) or rules (e.g., the rules of a model) to derive the answer (Episode C) (b) *A non-genuine mathematical argument in the absence of relevant established procedures or rules* (Episode D)	(c) To refute an answer to a problem involving a single calculation, find an error in the procedures or rules that were used to derive the answer (Episode C) (d) An argument by *reductio ad absurdum* (Episodes C, D)
Multiple but finitely many cases	(e) To justify that there are a specific number (finite) of possibilities involved in a situation, consider systematically all of the possible cases (Episodes A, B, E, F) (f) *An empirical argument based on non-systematic consideration of some (or even all) of the possible cases involved in a situation* (Episodes A, E)	(g) To refute a claim that there are a specific number (finite) of possibilities involved in a situation, show that their actual number is different (Episodes B, F) (h) An argument by *reductio ad absurdum* (Episodes A, E)
Infinitely many cases	(i) To justify that a set has an infinite cardinality, show that a subset of it has infinitely many elements that can be generated by a certain procedure which can run *ad infinitum* (Episodes A, B) (j) A generic argument (Episode G) (k) To justify a generalization over an infinite set, deduce logically the generalization by using definitions or other accepted statements (Episodes G, H) (l) *An empirical argument based on consideration of a proper subset of all of the possible cases in the domain of a generalization over an infinite set* (Episode H)	(m) An argument by *reductio ad absurdum* (Episode H) (n) To refute a generalization over an infinite set, find a counterexample to the generalization (Episode G) (o) *To refute a generalization over an infinite set, find multiple counterexamples to the generalization* (Episode G)

have used both of them in my work, I question whether the analyses could have produced by themselves trustworthy entries in all of the cells in the table. In particular, the mathematical analysis would have had little basis on which to determine the capacity or potential of elementary students to produce some but not other arguments from all of the viable alternatives, including possible differences in the use of related arguments such as *proof by contradiction* (which has no occurrence in Table 8.1) and *reductio ad absurdum* (which has multiple occurrences).[1] Also, as I have explained in Chapter 6 and elsewhere in the book, mathematics education research on different kinds of student arguments has tended to focus on post-elementary education and has placed uneven emphasis on the different kinds of proving tasks that I have identified in the book—proving tasks involving an infinite number of cases and with the purpose of justification have dominated research attention in this area (Stylianides et al., 2016b). Accordingly, an analysis of actual classroom work is also important for an empirically grounded and pedagogically relevant investigation of the relationship between proving tasks and proving activity in the elementary classroom. Indeed, the arguments in Table 8.1 are neither pure mathematical possibilities nor extrapolations from related but mostly distant research findings; rather, they are distilled versions of arguments offered or implied during actual mathematical work in two elementary classes, as described in the eight episodes.

I now consider the arguments in Table 8.1 across the following two dimensions of variation of the proving tasks where they were offered or implied: (1) variation in the *purpose* of a proving task (justification or refutation); (2) variation in the *number of cases* involved in a proving task (a single case, multiple but finitely many cases, or infinitely many cases). I shall consider each dimension separately.

With regard to the first dimension, I observe that there is little if any overlap in the arguments *across* the two columns of the table, which means that the students generally engaged in qualitatively different proving activities as they aimed to justify or to refute different statements. This observation is in agreement with the contrasting mathematical natures of justification and refutation, which are associated with statements of opposite truth-values (true and false, respectively).

With regard to the second dimension, I observe again that there is little overlap in the arguments *across* the three rows of the table that correspond to the kinds of proving tasks involving a single case, multiple but finitely many cases, and infinitely many cases. With the exception of *reductio ad absurdum*, which was used in all three of these kinds of tasks (items d, h, and m in Table 8.1), each of the other arguments appeared to be associated with only one kind. The empirical argument (items f and l) could appear at first sight to be associated with two kinds of tasks, but even this argument took two rather different forms where it was used: In one of these forms the inconclusive evidence that the empirical argument offered for the truth of a statement derived from the *non-systematic* consideration of cases in a finite set (item f), whereas in the other form it derived from the *incomplete* consideration of cases in an infinite set (item l).

Thus, we may conclude, on the basis of the main arguments offered in the eight episodes, that variations in the *purpose* of proving tasks or in the *number of cases* involved

[1] Definitions of *proof by contradiction* and *reductio ad absurdum* were offered in Table 2.1.

in them can support (independently or in combination) qualitatively different proving activities during the implementation of the tasks in the elementary classroom. A pedagogical implication of this conclusion is that some types of proving tasks (i.e., proving tasks whose characteristics differ across the two dimensions of variation) can be more appropriate than others to serve specific learning goals. For example, learning goals related to generic arguments (item j in Table 8.1) can be better served by proving tasks involving *justification* of a generalization over an *infinite* set than by other kinds of proving tasks.

An issue that is raised, then, is whether one of the dimensions of variation might be, from a pedagogical point of view, more suitable than the other to serve as an organizer of the proving tasks used in an elementary classroom. I propose that organizing proving tasks according to the number of cases involved in them might be more sensible pedagogically than organizing them according to their purpose. There are two main reasons for this. First, if proving tasks are allowed to evolve through a process of conjecture and revision, as often happened in the episodes in the context of students' engagement in mathematics as a sense-making activity, then the purposes of justification and refutation can both come into play within the context of the same proving task. Second, as suggested by an analysis of the arguments in Table 8.1, there are some commonalities in the mathematical resources used, or called for use, by students as they develop justification or refutation arguments within the same kind of proving task involving a certain number of cases. For example, arguments (a) and (c) both entail use of procedures or rules to derive the answer to a calculation problem; arguments (e) and (g) both entail consideration of a list of possibilities; and arguments (k) and (n) both entail appreciation of the notion of generality in a statement.

Variations in the number of cases involved in proving tasks can happen across lessons as part of the mathematical work of a class, where each lesson or sequence of related lessons uses the same kind of proving task (with respect to the number of cases involved in a task). This is essentially what happened in Episodes C–H (Chapters 5–7). Yet variations in the number of cases involved in proving tasks can also happen within a single lesson that uses of a special kind of proving task: tasks with *ambiguous conditions*. As I discussed in the context of Episodes A and B (Chapter 4), ambiguities in the conditions of a proving task can trigger or enable variation in students' perceptions of the number of cases involved in the task (Episode B) or variation in the actual number of cases involved in it as the teacher clarifies the task's conditions during its implementation in the classroom (Episode A). Accordingly, a proving task with ambiguous conditions can generate proving activity that spans sets of different cardinalities; this is reflected also in the spread of the references to Episodes A and B in Table 8.1.

Of course it is not enough to simply observe that different kinds of proving tasks can generate qualitatively different proving activities during their implementation in the elementary classroom. It is equally, if not more, important to describe the nuances of each form of proving activity, notably the student arguments offered during the activity, and to try to understand how the characteristics of the respective proving tasks shaped or influenced that activity (given also the mediating role of the teacher). In Chapters 5–7, I discussed these issues separately for each kind of proving task involving a different number of cases. However, one important issue I did not address, and which I mention as a direction for future research, is how to organize more systematically the different arguments (both invalid and valid) that elementary students can offer as they engage with each different kind of proving task. This can result, for example, in three separate categorizations of student arguments, ranging from invalid arguments to arguments that meet the standard of proof. The arguments in those categorizations (essentially hierarchies) could find application as points in possible "learning trajectories" (Simon, 1995) that elementary students would be supported to follow during their engagement with the respective kinds of proving tasks. The arguments in Table 8.1 could be a useful starting point for the development of those categorizations, and could be used alongside existing categorizations of student arguments (e.g., Balacheff, 1988a; Harel & Sowder, 2007; Marrades & Gutiérrez, 2000; Simon & Blume, 1996). The existing categorizations, though, have paid uneven attention to the various kinds of proving tasks that I have considered in this book, and so they may be more relevant to some of them than to others.

Finally, I acknowledge that the aspects of the relationship between proving tasks and proving activity that I have articulated in this book might have been influenced by the decision I made to focus on specific categories of problems within each kind of proving task. In particular, in my examination of proving tasks involving a single case in Chapter 5, I focused on problems that are placed in the context of a single calculation. Also, in my examination of proving tasks involving multiple but finitely many cases in Chapter 6, I focused on combination, permutation, and Cartesian product problems that ask students to find all possibilities involved in a situation and justify that they have found all of them. While not comprehensive, the specific categories of problems I have focused on relate to a large part of the mathematical work in elementary schools. My decision in Chapter 5 in particular to focus on calculation problems may be seen as a strategic investment of research effort: Given how large a part calculation work occupies in the elementary mathematics curriculum, and also the powerful aspect of mathematical sense-making that proving can bring to this work (see discussion of Episodes C and D), one may talk about a transformative function that an elevated place of proving in calculation work can play in elementary school mathematics.

The Role of the Teacher

My discussion of all eight episodes highlighted the important role of the teacher in mediating the relationship between proving tasks and corresponding proving activity. Indeed, the actions performed by a teacher during the selection or design of a mathematics task, as well as during the implementation of the task in the classroom, constitute a major influence on the opportunities offered to students to engage in mathematical work (e.g., Christiansen & Walther, 1986; Henningsen & Stein, 1997; Sullivan et al., 2013), proving activity being a case in point (e.g., Sears & Chávez, 2014). In what follows I discuss aspects of the teacher's role in *selecting or designing proving tasks* and in *implementing proving tasks* in the elementary classroom.

Selecting or Designing Proving Tasks

With regard to task selection or design, the information in Table 8.1 shows clearly that unless elementary teachers use proving tasks across the whole range of possibilities their students will only have opportunities to engage with certain kinds of proving activity and will thus be deprived of access to broad-based learning in the area of proving. For example, the students of an elementary teacher who does not generally use proving tasks involving true or false generalizations over infinite sets are essentially denied opportunities to learn about certain kinds of arguments that are particularly relevant to these tasks, such as generic arguments (item j in Table 8.1) and refutation by counterexample (item n). But the consequences for students' learning can be more serious than just the lack of learning opportunities: The exclusive use or the overemphasis of some kinds of proving tasks at the expense of others can also cultivate in students unhelpful conceptions about what it means to prove more generally and can thus create obstacles in their (future) work with proving tasks that call for qualitatively different proving activities. This position has a parallel in the curricular area of whole-number arithmetic, where it was found that the imbalanced focus on learning experiences that promote certain meanings of the basic operations at the expense of other meanings creates obstacles in students' learning of those other meanings (e.g., Carpenter et al., 1996; Fischbein et al., 1985).

Consider, for example, an elementary teacher who uses predominantly proving tasks involving multiple but finitely many cases. The students' experiences with these tasks are likely to cultivate in them a limited conception of proof as an argument that considers (systematically or not) cases involved in a statement (items e and f in Table 8.1); this conception is unhelpful when it comes to justifying statements over infinite sets. Indeed, Episode H (Chapter 7) offered evidence to suggest that students' empirical arguments for generalizations over infinite sets (item l) may derive, at least partly, from their attempts to transfer to the domain of infinite sets proving strategies that are generally applicable only in finite sets.

Implementing Proving Tasks

With regard to task implementation, the teachers' actions in the eight episodes that I examined in Chapters 4–7 can be clustered into three broad categories. While this categorization is necessarily selective and runs the risk of masking the nuances of the pedagogical practices observed in individual episodes, it nevertheless highlights some general patterns in the two teachers' actions, thus offering some insights into the role of the teacher more broadly.

Category 1: setting up the proving tasks

The first category of teacher actions related to *setting up the proving tasks*, and it often involved the teachers engaging their students in an initial exploration of a mathematical situation before raising explicitly the issue of proof. The initial exploration helped motivate the need for a proof, which was mainly used as a means to justify or refute a conjecture (e.g., Episode A, Chapter 4, and Episodes G and H, both Chapter 7) but also to reconcile a disagreement (e.g., Episodes B and D, Chapters 4 and 5, respectively) or to convince a skeptic (Episode

E, Chapter 6). This way of infusing the issue of proof into students' mathematical activity is consistent with the perspective of proving as a vehicle to mathematical sense-making that I have adopted in this book and that has also been elaborated in the literature (e.g., Ball & Bass, 2000b, 2003; Hanna, 1990, 1995).

Interestingly, though, in some proving tasks where the issue of proof was not explicitly raised by the teacher as part of the set-up of the task, that issue seemed nevertheless to be understood by the class as an unstated but clear expectation of their work on the task (e.g., Episode A, Chapter 4). The shared understanding of expectations regarding proof and its role in the mathematical work of a class may be attributed to the "sociomathematical norms" in the class (Yackel & Cobb, 1996). Indeed, once the students in a class have been inducted to a culture of argumentation and proof, the issue of motivating the need for proof, which can be critical at the beginning of that process (Episode E, Chapter 6), may not be an explicit concern of the teacher during the set-up of every proving task. As also illustrated by the episode from Zack's (1997) fifth-grade class that I presented in Chapter 1, once classroom norms that support argumentation and proof have been established, students themselves can also raise the issue of proof and can engage in proving activity independently of the teacher's presence. The teacher can scaffold the development and stabilization of this kind of classroom norm in an elementary classroom over a rather long period of time (Makar, Bakker, & Ben-Zvi, 2015).

Category 2: managing students' solutions to the proving tasks

The second category of teacher actions related to *managing students' solutions to the proving tasks,* and it involved use of a range of practices elaborated in the literature, especially in studies that examined teachers' orchestration of whole class discussions in the context of cognitively demanding tasks. Some main practices that the two teachers used, to varying degrees but quite routinely, in virtually all of the episodes are the following: asking questions and eliciting students' ideas, allowing also time and opportunities for students to formulate and communicate their ideas (e.g., Cengiz, Kline, & Grant, 2011; Kazemi et al., 2009; Nicol, 1998; Stein et al., 2008; van Es & Sherin, 2002); encouraging students to listen and respond to each other's ideas, identifying also any key points of disagreement in their contributions and turning some of them into the subject of productive discussions (e.g., Reid, 2002; Wood, 1999; Zack, 1997); filtering students' ideas and highlighting for attention (e.g., through re-voicing), or selecting for discussion, those ideas that are mathematically important and particularly relevant to the task at hand (e.g., O'Connor & Michaels, 1996; Sherin, 2002; Stein et al., 2008; Stylianides & Stylianides, 2014b).

These are rather generic practices that tend to characterize competent mathematics teaching in different areas (including proving) as well as competent teaching more generally. The use of such practices in the course of teaching tends to place high demands on teachers for improvisation and in-the-moment decision making (e.g., Scherrer & Stein, 2013; Stein et al., 2008; van Es & Sherin, 2002), and indeed this is what happened at several points during the episodes. Take, for example, Ball's decision in Episode H (Chapter 7) to pick up on Jeannie and Sheena's idea regarding the provability of conjectures over infinite sets and to organize subsequent classwork around it, or Howard's decision in Episode G (Chapter 7) to pick up on Jack's generic argument and make it an explicit focus of discussion with the rest of the class. In both of these cases, the teachers contingently responded to important classroom events by *filtering* students' ideas and *selecting* for attention and discussion those ideas that they considered to be, and indeed were, key to the mathematical issues at hand and to the advancement of students' learning.

Notwithstanding the high demands that the use of such practices can place on teachers to act in the situation, thus appearing to suggest that only "expert" teachers would be able to use them effectively in their teaching, recent research developments suggest that these practices can be learnable (Kazemi et al., 2009; Stein et al., 2008; van Es & Sherin, 2002) and that their improvisational use can be downplayed in favor of advance, empirically tested, and well-theorized instructional planning (Stylianides & Stylianides, 2014b). Research is needed to understand how the effective use of these practices in the teaching of proving can be made manageable for more elementary teachers including "non-experts."

Category 3: helping students overcome hurdles in their proving activity

The third category of teacher actions related to *helping students overcome hurdles in their proving activity*, and it involved a range of proving-specific interventions that the teachers undertook to "scaffold" students' work (cf. Bakker et al., 2015; Maybin et al., 1992; Wood et al., 1976) and to connect students with conventional mathematical knowledge that they might not have been able to access otherwise (cf. Ball, 1993; Lampert, 1992; Stylianides, 2007b; Yackel & Cobb, 1996). Each intervention targeted, in the given classroom situation, the specific component or components of a mathematical argument that, according to my own analysis based on the definition of proof used in this book (see Chapter 2), caused difficulties for students' proving activity or created obstacles to their production of a proof. Indeed, while the two teachers were unaware of this definition and of the three components into which it breaks down any given mathematical argument—the set of accepted statements, the modes of argumentation, and the modes of argument representation (Stylianides, 2007b)—the teachers' interventions turned out to be consistent with the definition in terms, for example, of what component or components of a non-proof argument needed to be developed so that the argument would meet the standard of proof. In what follows I describe examples of interventions that the teachers used in different episodes, each time targeting a different component of an argument.

In Episode D (Chapter 5) students' proving activity on the problem $6 + \hat{6} = ?$ came up against a hurdle due to the lack of the concept of addition of a negative integer in the community's *set of accepted statements*. As a result of this lack of necessary mathematical knowledge, some students ended up offering non-genuine mathematical arguments to justify their answers to the problem (item b in Table 8.1). For example, Sean said the answer would be 6 because the term $+ \hat{6}$ "would just disappear." Ball's intervention targeted the community's set of accepted statements: She set out to expand this set by introducing the students to a new model for addition that could help them develop their concept of integer addition.

In Episode F (Chapter 6) the proving activity of the students in Howard's class came up against a hurdle as the students tried to manage the representational complexity involved in listing systematically all possible cases in a finite set (item e in Table 8.1). Howard's intervention targeted the *modes of argument representation* used in the class: She helped the students recognize the limitations of the modes of representation they had been using (realistic drawings of outfits) and introduced them to an efficient mode of representation (a "code" that used a letter or a digit for each part of the outfit). This efficient mode empowered several students to solve proving tasks with high representational complexity (e.g., the Further Extended Outfits Problem). Another example of an intervention that targeted the *modes of argument representation* used by students is found in Episode G (Chapter 7), where Howard took the leading role in guiding students through the development of Jack's generic argument (item j in Table 8.1) into a more general proof (item k) using algebraic notation.

Examples of interventions that targeted students' *modes of argumentation* can be found in Episodes E (Chapter 6), H, and G (both Chapter 7). In Episode E the students in Ball's class were relying on empirical arguments to justify that all possibilities in a finite set have been found (item f in Table 8.1). Ball's intervention comprised a firm stance of disapproval of students' empirical arguments and acting in the role of a "skeptic" (Mason, 1982), with a persistent critique of the leap of faith that students' arguments were asking one to make. At the end, a student in the class, Betsy, offered an argument that used a valid mode of argumentation and met the standard of proof (item e). Encouragingly, the class accepted Betsy's argument as a proof with Lisa, for example, calling it "a really neat idea."

In Episode H the students in Ball's class came up against another hurdle as they worked on a true conjecture over an infinite set—whether and how one could prove statements involving infinitely many cases—with some students offering empirical arguments for the conjecture (item l in Table 8.1). Ball's intervention built on an idea mentioned by Jeannie and Sheena about the provability of statements over infinite sets, which captured the essence of the hurdle. The intervention targeted the *modes of argumentation* required for a proof, and it comprised introducing students to the way in which mathematicians use definitions to deduce logically statements such as the conjecture at hand. As happened in Episode E, Betsy ended up offering an argument that met the standard of proof (item k). However, this time Betsy's argument received opposition and disbelief from the class, which suggests that the intervention might have not impacted the thinking of many students as it had done for Betsy.

In Episode G the students in Howard's class were resistant to accepting that a single counterexample sufficed to refute a false conjecture over an infinite set (item n in Table 8.1), and they required multiple counterexamples for that purpose (item o). Howard's intervention targeted the *modes of argumentation* required for a proof (refutation). The intervention comprised persistent questioning of students' ideas about the number of counterexamples they thought were required to refute a generalization, an issue that Howard brought up repeatedly and in the context of several false conjectures over infinite sets. However, the students did not seem to accept the valid mode of argumentation promoted by Howard, which suggests that the intervention might not have had the intended impact on students' thinking. The class ended up working on a conjecture that, although originally false, was subsequently revised by the class to be true, thus making the issue of refutation no longer applicable.

The interventions I have described had varying degrees of success. In particular, the interventions in Episodes H and G, which targeted *modes of argumentation* used in justifying or refuting generalizations over infinite sets, had a questionable impact on the thinking of the class in which they were implemented. This is not surprising, as the two teachers were navigating with these interventions pedagogically arduous territories that are also relatively uncharted by research, especially at the elementary school level (Stylianides et al., 2016b). Furthermore, in these two episodes, as in almost all other episodes where the two teachers carried out an intervention to help students overcome hurdles in their proving activity, the teachers were acting *in the situation*, improvising on the design of the interventions and contingently responding to students' contributions as the interventions unfolded. For example, the intervention in Episode H was contingent on the idea that was expressed by Jeannie and Sheena regarding the provability of generalizations over infinite sets. Also, the interventions in Episodes E and H were contingent on the arguments (proofs) that were developed by Betsy and Jack, respectively. What might have happened if neither these nor other students had made these contributions? In such a case, the teachers might have had to come up, in real time, with alternative interventions or with modifications to the ones they were in the process of implementing.

If helping elementary students overcome hurdles in their proving activity is left to be a didactical problem of improvisation and in-the-moment decision making, only a few elementary teachers will be willing to take up the challenge and even fewer of them will be able to solve the problem adequately. This points out a pressing need for research to develop effective interventions that elementary teachers can use, or adapt for use, to help their students overcome the various hurdles encountered in their proving activity, especially those relating to modes of argumentation for justifying or refuting generalizations over infinite sets. While there is a scarcity of research on this topic at the elementary school level, there are some promising findings from relevant intervention studies with older students that can inform future research at the elementary school level (Anderson et al., 1995; Brown, 2014; Harel, 2002; Hodds et al., 2014; Jahnke & Wambach, 2013; Mariotti, 2000, 2013; Schoenfeld, 1985; Stylianides & Stylianides, 2009b; Weber, 2006). The design of such interventions can also benefit from work on the role of instructional engineering in reducing the demands on the teacher for improvisation and in-the-moment decision making during the implementation of an intervention in the area of proving (Stylianides & Stylianides, 2014b), as well as from more general accounts of the critical role that the duration of an intervention has in its potential to be adopted by teachers or to be implemented as intended (Stylianides & Stylianides, 2013, 2014a; Yeager & Walton, 2011).

The Place of Proving in Elementary Students' Mathematical Work

In this section I revisit the book's broad aim and discuss implications of the work reported herein for what it might take to elevate the place of proving in elementary students' mathematical work. My discussion is organized around two themes, *teacher education* and *curricular resources*, and aims to address in a rather holistic way the four factors that I identified in Chapter 2 as potentially important contributors to the currently marginal place of proving in the elementary mathematics classroom.

Specifically, these factors related to the weak mathematical knowledge that many elementary teachers have about proof (factor 1) and their presumed beliefs that proving is an advanced mathematical topic beyond the reach of elementary students (factor 2); the high pedagogical demands placed on elementary teachers who strive to engage their students in proving (factor 3); and the inadequate instructional support offered or available to elementary teachers about how to achieve that goal in their classrooms (factor 4). As I noted in Chapter 2, factors 1 and 2 are not meant to support a deficit model of teacher competence but rather to emphasize, alongside factor 3, the challenges involved for elementary teachers in undertaking the demanding endeavor to engage young children in proving, a hard-to-teach and hard-to-learn mathematical activity. Accordingly, the locus of responsibility shifts from factors 1–3, which concern mostly elementary teachers, to factor 4, which concerns several stakeholders, such as teacher educators and curriculum developers (including textbook authors), but also researchers who could aim to provide the relevant research basis for the required instructional support.

Teacher Education

Even if teacher educators recognize, as I presume they do, the importance of proving in elementary school mathematics, there appears to be a scarcity of research or resources that they can draw on in preparing elementary teachers (pre-service or in-service) to engage their students in proving (McCrory & Stylianides, 2014; Stylianides et al., 2016a, b). The work reported in this book can help address this gap, thus increasing the chances that, through better preparation in teacher education, elementary teachers will be better able, or more willing, to give proving an elevated place in their students' mathematical work.

Teacher educators can use this book in different ways. In particular, they can use it as a resource for themselves to identify key issues related to proving tasks, proving activity, and the teacher's role, and offer a distilled version of main insights, in the form of guidance, to elementary teachers. Another possible use of the book (or individual chapters thereof) is as an assigned reading, which can precede or follow a practical exploration related to proving that teachers can be asked to carry out with elementary students in the classroom. This practical exploration could ask teachers to select or design a proving task, implement it in their classrooms (regular classrooms if in-service teachers, or field placement classrooms if pre-service teachers), and reflect both on the students' proving activity and on their own role in managing that activity.

The discussion earlier in this chapter makes it clear that the choice of a proving task will influence the issues likely to arise for teachers, including the challenges they may face in getting students to develop a proof, the actions they may have to take to help students overcome hurdles in their proving activity, etc. This knowledge can therefore be an important consideration for teacher educators as they specify the parameters of the exploration. For example, if teacher educators impose no restrictions on the kinds of proving tasks chosen by the teachers, they should expect that those teachers who, say, choose tasks involving justification or refutation of generalizations over infinite sets will have a more challenging experience than those who choose tasks involving justification of a single case. This is not necessarily a problem, but it is nevertheless an important factor to consider alongside the teacher educators' particular goals for the exploration. Furthermore, the specific prompts that the teacher educators may design for teachers' reflection on their exploration can help address issues related to any of the four factors described at the beginning of this section. Specifically, prompts about mathematical issues involved in the exploration can address issues related to factor 1; prompts about pedagogical challenges emerging from the exploration can address issues related to factor 3; and so forth.

Teacher educators can also use the book (or individual chapters thereof) during sessions in teacher education. In particular, sessions can be organized around the descriptions of one or more classroom episodes from the book, which can be viewed as "narrative cases of teaching practice" and can be used to raise a multitude of mathematical or pedagogical issues for reflection and discussion during the sessions (cf. Smith, Silver, Stein, Boston, & Henningsen, 2005; Stein, Smith, Henningsen, & Silver, 2000). Depending again on the goals of the teacher educators, sessions can address issues related to any of the four factors. For example, the participants in these sessions can be asked to analyze mathematically the proving tasks and respective student arguments described in any pair of episodes from Chapters 4–7 (cf. factor 1). They can also be asked to reflect on whether and why the mathematical work described in the episodes is appropriate for or beneficial to the students, thus helping bring to the fore participants' own beliefs about proving as a legitimate curricular topic in elementary school mathematics (cf. factor 2). They can be asked further to analyze the teachers' actions described in the episodes, including possible dilemmas or challenges that the teachers might have faced as they listened to students' arguments, orchestrated whole class discussions around students' solutions to a proving task, and so forth (cf. factor 3). Yet another possibility is to ask the participants to analyze the overall instructional plan that seemed to have underpinned the teachers' actions in the episodes, including the teachers' choices of proving tasks and the student learning goals they aimed to accomplish (cf. factor 4). In facilitating these sessions, the teacher educators can draw on the discussion of the episodes in the book, as well as on other relevant parts of the book, to identify key points and use them during the session to steer progression toward the goals of the sessions. Some of these parts of the book can be assigned, then, as post-session readings.

I clarify that, no matter how relevant or useful the book can be to teacher education, it is not meant to be an independent (or the sole) basis of preparation for elementary teachers wanting to engage their students in proving. For example, teacher educators may have to first help elementary teachers appreciate a need for proof (Stylianides & Stylianides, 2015; Zaslavsky et al., 2012) before they can expect elementary teachers to productively engage with the mathematical or pedagogical nuances of engaging elementary students in proving as described in this book.

Curricular Resources

Curricular resources in general, and mathematics textbooks in particular, play, or can play, an important role in many teachers' everyday practice and, by implication, in the learning opportunities offered to students to engage in mathematical activity including proving (e.g., Bieda, 2010; Cai et al., 2011; Stylianides, 2016). While there are countries like England where reliance on textbooks is considerably lower than in other countries, TIMSS 2011 (Mullis et al., 2012) showed that textbooks are the most frequent basis of mathematics instruction at both fourth and eighth grades, used on average with approximately 75% of students internationally. However, there appears to be a scarcity of mathematics textbooks that adequately address issues of proving at the elementary level (Bieda et al., 2014) and beyond (Davis et al., 2014; Fujita & Jones, 2014; Otten et al., 2014; Stylianides, 2008b, 2009b; Thompson et al., 2012). The work reported in this book can inform the design of such textbooks, thus increasing the chances that, through the use of these textbooks, elementary teachers will be better able or more willing to give to proving an elevated place in their students' mathematical work.

Specifically, this book has identified a range of proving tasks that can engage elementary students in productive proving activity and whose balanced representation in elementary mathematics textbooks can support a rounded set of learning experiences for students in the area of proving. In the section "Selecting or Designing Proving Tasks" earlier in this chapter, I explained the important ramifications that an imbalanced use of different kinds of proving tasks can have for students' learning in this area. However, it is a complex empirical question what might be an appropriate way for textbooks to allocate across the school years the learning experiences in the area of proving that textbooks design for students (Stylianides, 2009b).

In addition to providing instructional support to elementary teachers in the form of proving tasks they can use in their classrooms (cf. factor 4), textbooks can offer teachers other forms of guidance. In an accompanying teacher's edition, textbook authors can draw, for example, on the analysis of the relationship between proving tasks and proving activity elaborated in this book to help teachers anticipate what arguments (both valid and invalid) their students might offer in response to a given proving task (cf. factor 3). A mathematical analysis of the different arguments (their limitations if invalid, their strengths if proofs) can also help support teachers' own mathematical knowledge (cf. factor 1). Furthermore, the very inclusion of proving tasks in elementary mathematics textbooks, together with the relevant guidance for their implementation in the accompanying teacher's edition, may help elevate, in the eyes of elementary teachers, the value of proving in their students' mathematical work (cf. factor 3).

Those textbooks and their accompanying teacher's edition that aim together to support not only student learning but also teacher learning, by offering the forms of instructional support and guidance I have described, are often referred to in the literature as "educative curriculum materials" (Ball & Cohen, 1996; Davis & Krajcik, 2005; Davis et al., 2014; Stylianides, 2008b, 2014). Educative curriculum materials can play a crucial role in effective instruction in general, but especially in the instruction of activities such as proving that are recognized to be hard for teachers to teach and hard for students to learn (Stylianides & Stylianides, 2014b). Of course, following the guidance offered in educative curriculum materials might be a non-trivial accomplishment for many teachers (Remillard, 2005), which relates to the broader issue of the complex interplay between curricular resources and their classroom use (Stylianides, 2016). Yet in those countries where textbooks play a major role in teachers' everyday practice (see Mullis et al., 2012), educative curriculum materials can reach a large number of teachers and serve potentially as vehicles for instructional reform (Ball & Cohen, 1996), with the area of proving in the elementary mathematics classroom being a potential case in point.

Epilogue

Elementary school mathematics tends to be associated primarily with number facts, calculations, and algorithms. These are important curricular topics and essential for students' mathematical education, but they are not in themselves "authentic mathematics" (cf. Lampert, 1992). Indeed, one could call that work "numeracy" rather than "mathematics." What is fundamentally missing from that work is an appreciation of what counts as evidence in mathematics, i.e., the grounds that support the conclusions one draws and the means by which these conclusions are derived and represented. The activity of proving lies at the heart of the standards of evidence in mathematics and is indispensable to any work that can genuinely be called mathematical. Thus the fact that proving is typically not a part of elementary students' mathematical work is a serious threat to the integrity of their learning experiences and a problem in urgent need of solution.

In this book I took a step toward addressing this problem by examining how the place of proving in elementary students' mathematical work can be elevated through the purposeful design and implementation of mathematics tasks, specifically proving tasks. In doing so, I cast some light on the relationship between different kinds of proving tasks and the proving activity they can help generate in the elementary classroom, and I have articulated aspects of the role of elementary teachers in mediating that relationship. The work reported in this book has implications for teaching, teacher education, and curricular resources, and makes a contribution toward the realization (in the long term) of an ambitious vision for elementary school mathematics whereby classroom episodes like those described in this book will be considered the norm rather than the exception.

REFERENCES

Alcock, L. (2010). Mathematicians' perspectives on the teaching and learning of proof. In: F. Hitt, D. Holton, & P. Thompson (eds.), *Research in collegiate mathematics education VII* (pp. 63–92). Providence, RI: American Mathematical Society.

Alcock, L. & Simpson, A. P. (2004). Convergence of sequences and series: Interactions between visual reasoning and the learner's beliefs about their own role. *Educational Studies in Mathematics*, **57**, 1–32.

Alcock, L. & Weber, K. (2010). Referential and syntactic approaches to proving: Case studies from a transition-to-proof course. *Research in Collegiate Mathematics Education*, **7**, 101–123.

Alexander, R. J. (2006). *Towards dialogic teaching* (3rd edn.). New York, NY: Dialogos.

Anderson, J. A., Corbett, A., Koedinger, K., & Pelletier, R. (1995). Cognitive tutors: Lessons learned. *Journal of the Learning Sciences*, **4**, 167–207.

Arsac, G. & Durand-Guerrier, V. (2005). An epistemological and didactic study of a specific calculus reasoning rule. *Educational Studies in Mathematics*, **60**, 149–172.

Arzarello, F., Andriano, V., Olivero, F., & Robutti, O. (1998). Abduction and conjecturing in mathematics. *Philosophica*, **61**(1), 77–94.

Bakker, A., Smit, J., & Wegerif, R. (2015). Scaffolding and dialogic teaching in mathematics education: introduction and review. *ZDM—The International Journal on Mathematics Education*, **47**, 1047–1065.

Balacheff, N. (1988a). Aspects of proof in pupils' practice of school mathematics. In: D. Pimm (ed.), *Mathematics, teachers and children* (pp. 216–235). London: Hodder & Stoughton.

Balacheff, N. (1988b). A study of students' proving processes at the junior high school level. In: I. Wirszup & R. Streit (eds.), *Proceedings of the Second UCSMP International Conference on Mathematics Education* (pp. 284–297). Reston, VA: National Council of Teachers of Mathematics.

Balacheff, N. (1990). Towards a *problématique* for research on mathematics teaching. *Journal for Research in Mathematics Education*, **21**, 258–272.

Balacheff, N. (1991). The benefits and limits of social interaction: The case of mathematical proof. In: A. J. Bishop (ed.), *Mathematical knowledge: Its growth through teaching* (pp. 175–192). Dordrecht, Netherlands: Kluwer Academic Publishers.

Balacheff, N. (2002). The researcher epistemology: a deadlock for educational research on proof. In: F. L. Lin (ed.), *Proceedings of the 2002 International Conference on Mathematics: understanding proving and proving to understand* (pp. 23–44). Taipei, Taiwan: NSC and NTNU. Pre-publication version retrieved November 25, 2011, from: http://www.tpp.umassd.edu/proofcolloquium07/reading/Balachef_Taiwan2002.pdf

Ball, D. L. (1993). With an eye on the mathematical horizon: Dilemmas of teaching elementary school mathematics. *The Elementary School Journal*, **93**(4), 373–397.

Ball, D. L. (2000). Working on the inside: Using one's own practice as a site for studying mathematics teaching and learning. In: A. Kelly & R. Lesh (eds.), *Handbook of research design in mathematics and science education* (pp. 365–402). Mahwah, NJ: Lawrence Erlbaum Associates.

Ball, D. L. & Bass, H. (2000a). Interweaving content and pedagogy in teaching and learning to teach: Knowing and using mathematics. In: J. Boaler (ed.), *Multiple perspectives on the teaching and learning of mathematics* (pp. 83–104). Westport, CT: Ablex.

Ball, D. L. & Bass, H. (2000b). Making believe: The collective construction of public mathematical knowledge in the elementary classroom. In: D. Philips (ed.), *Constructivism in education: Yearbook of the National Society for the Study of Education* (pp. 193–224). Chicago, IL: University of Chicago Press.

Ball, D. L. & Bass, H. (2003). Making mathematics reasonable in school. In: J. Kilpatrick, W. G. Martin, & D. Schifter (eds.), *A research companion to principles and standards for school mathematics* (pp. 27–44). Reston, VA: National Council of Teachers of Mathematics.

Ball, D. L. & Cohen, D. K. (1996). Reform by the book: What is—or might be—the role of curriculum materials in teacher learning and instructional reform? *Educational Researcher*, **25**(9), 6–8.

Ball, D. L., Hoyles, C., Jahnke, H. N., & Movshovitz-Hadar, N. (2002). The teaching of proof. In: L. I. Tatsien (ed.), *Proceedings of the International Congress of Mathematicians*, Vol. III (pp. 907–920). Beijing: Higher Education Press.

Bartolini Bussi, M. G. (2000). Early approach to mathematical ideas related to proof making. In: P. Boero, P., G. Harel, C. Maher, & M. Miyazaki (eds.), *Proof and proving in mathematics education. Proceedings of the ICME9—TSG12*. Tokyo: Makuhari. Retrieved July 3, 2013, from http://www.lettredelapreuve.it/OldPreuve/ICME9TG12/ICME9TG12Contributions/BartoliniBussiICME00.html

Bell, A. W. (1976). A study of pupil's proof-explanations in mathematical situations. *Educational Studies in Mathematics*, 7(1), 23–40.

Bieda, K. N. (2010). Enacting proof-related tasks in middle school mathematics: challenges and opportunities. *Journal for Research in Mathematics Education*, **41**, 351–382.

Bieda, K. N., Ji, X., Drwencke, J., & Picard, A. (2014). Reasoning-and-proving opportunities in elementary mathematics textbooks. *International Journal of Educational Research*, **64**, 71–80.

Blanton, M. L. (2008). *Algebra in elementary classrooms: Transforming thinking, transforming practice*. Portsmouth, NH: Heinemann.

Blanton, M. L. & Kaput, J. J. (2011). Functional thinking as a route into algebra in the elementary grades. In: J. Cai & E. Knuth (eds.), *Early algebraization: A global dialogue from multiple perspectives* (pp. 5–24). Heidelberg: Springer.

Boero, P., Garuti, R., & Mariotti, M. A. (1996). Some dynamic mental processes underlying producing and proving conjectures. *Proceedings of the 20th Conference of the International Group for the Psychology of Mathematics Education*, Vol. 2 (pp. 121–128). Valencia, Spain: Valencia University.

Boston, M. D. & Smith, M. S. (2009). Transforming secondary mathematics teaching: Increasing the cognitive demands of instructional tasks used in teachers' classrooms. *Journal for Research in Mathematics Education*, **40**(2), 119–156.

Britt, M. & Irwin, K. (2011). Algebraic thinking with and without algebraic representation. In: J. Cai & E. Knuth (eds.), *Early algebraization* (pp. 137–157). New York, NY: Springer.

Brousseau, G. (1997). *Theory of didactical situations in mathematics: Didactique des mathématiques 1970–1990* (translated and edited by N. Balacheff, M. Cooper, R. Sutherland, & V. Warfield). Dordrecht: Kluwer Academic Publishers.

Brown, S. (2014). On skepticism and its role in the development of proof in the classroom. *Educational Studies in Mathematics*, **86**, 311–335.

Bruner, J. (1960). *The process of education*. Cambridge, MA: Harvard University Press.

Buchbinder, O. & Zaslavsky, O. (2007). How to decide? Students' ways of determining the validity of mathematical statements. In: D. Pita-Fantasy & G. Philippot (eds.), *Proceedings of the 5th Congress of the European Society for Research in Mathematics Education* (pp. 561–571), Larnaca: University of Cyprus.

Cai, J. & Knuth, E. (2011). Introduction: A global dialogue about early algebraization from multiple perspectives. In: J. Cai & E. Knuth (eds.), *Early algebraization: A global dialogue from multiple perspectives* (pp. vii-xi). Heidelberg: Springer.

Cai, J., Ni, Y., & Lester, F. K. (2011). Curricular effect on the teaching and learning of mathematics: Findings from two longitudinal studies in China and the United States. *International Journal of Educational Research*, **50**, 63–64.

Carpenter, T. P., Fennema, E., & Franke, M. L. (1996). Cognitively guided instruction: A knowledge base for reform in primary mathematics instruction. *The Elementary School Journal*, **97**, 3–20.

Carpenter, T. P., Franke, M. L., & Levi, L. (2003). *Thinking mathematically: Integrating arithmetic & algebra in elementary school*. Portsmouth, NH: Heinemann.

Carpenter, T. P., Levi, L., Berman, P. W., & Pligge, M. (2005). Developing algebraic reasoning in the elementary school. In: T. A. Romberg, T. P. Carpenter, & F. Dremock (eds.), *Understanding mathematics and science matters* (pp. 81–98). Mahwah, NJ: Lawrence Erlbaum Associates.

Carraher, D. W., Schliemann, A. D., Brizuela, B. M., & Earnest, D. (2006). Arithmetic and algebra in early mathematics education. *Journal for Research in Mathematics Education*, **37**(2), 87–115.

Cengiz, N., Kline, K., & Grant, T.J. (2011). Extending students' mathematical thinking during whole-group discussions. *Journal of Mathematics Teacher Education*, **14**, 355–374.

Chazan, D. (1993). High school geometry students' justification for their views of empirical evidence and mathematical proof. *Educational Studies in Mathematics*, **24**(4), 359–387.

Christiansen, B. & Walther, G. (1986). Task and activity. In: B. Christiansen, A. G. Howson, & M. Otte (eds.), *Perspectives on mathematics education* (pp. 243–307). Dordrecht, Netherlands: Reidel Publishing Company.

Cobb, P., Yackel, E., & Wood, T. (1992). A constructivist alternative to the representational view of mind in mathematics education. *Journal for Research in Mathematics Education*, **23**, 2–33.

Coe, R. & Ruthven, K. (1994). Proof practices and constructs of advanced mathematics students. *British Educational Research Journal*, **20**, 41–53.

Collopy, R. (2003). Curriculum materials as a professional development tool: how a mathematics textbook affected two teachers' learning. *The Elementary School Journal*, **103**(3), 227–311.

Corey, D. & Gamoran, S. M. (2006). Practicing change: curriculum adaptation and teacher narrative in the context of mathematics education reform. *Curriculum Inquiry*, **36**(2), 153–187.

Davis, E. A. & Krajcik, J. S. (2005). Designing educative curriculum materials to promote teacher learning. *Educational Researcher*, **34**(3), 3–14.

Davis, E. A., Palincsar, A. S., & Arias, A. M. (2014). Designing educative curriculum materials: A theoretically and empirically driven process. *Harvard Educational Review*, **84**(1), 24–52.

Davis, J. D., Smith, D. O., & Roy, A. R. (2014). Reasoning-and-proving in algebra: The case of two reform-oriented U.S. textbooks. *International Journal of Educational Research*, **64**, 92–106.

Davis, P. & Hersh, R. (1981). *The mathematical experience*. New York: Viking Penguin Inc.

Davis, R. B. (1992). Understanding "understanding." *Journal of Mathematical Behavior*, **11**, 225–241.

Davis, R. B. & McKnight, C. (1976). Conceptual, heuristic, and S-algorithmic approaches in mathematics teaching. *Journal of Children's Mathematical Behavior*, **1**, 271–286.

de Millo, R., Lipton, R., & Perlis A. (1979/1998). Social processes and proofs of theorems and programs. In: T. Tymoczko (ed.), *New directions in the philosophy of mathematics* (pp. 267–285). Princeton, NJ: Princeton University Press.

Department for Education (2013). *Mathematics programmes of study: Key stages 1–2* [National Curriculum in England]. Retrieved October 31, 2013, from: https://www.gov.uk/government/uploads/system/uploads/attachment_data/file/239129/PRIMARY_national_curriculum_-_Mathematics.pdf de Villiers, M. (1986). *The role of axiomatisation in mathematics and mathematics teaching*. University of Stellenbosch, South Africa. Retrieved November 20, 2015, from: https://www.researchgate.net/profile/Michael_Villiers/publication/252290335_The_Role_of_Axiomatisation_in_Mathematics_and_Mathematics_Teaching/links/53e13be50cf2d79877a59c95.pdf

de Villiers, M. (1990). The role and function of proof in mathematics. *Pythagoras*, **24**, 17–24.

de Villiers, M. (1994). The role and function of a hierarchical classification of quadrilaterals. *For the Learning of Mathematics*, **14**(1), 11–18.

de Villiers, M. (1999). The role and function of proof. In: M. de Villiers (ed.), *Rethinking proof with the Geometer's Sketchpad* (pp. 3–10). Emeryville, CA: Key Curriculum Press.

Dewey, J. (1902). *The child and the curriculum*. Chicago, IL: University of Chicago Press.

Dewey, J. (1903). The psychological and the logical in teaching geometry. *Educational Review*, **XXV**, 387–399.

Dewey, J. (1910/1997). *How we think*. Mineola, NY: Dover Publications.

Doyle, W. (1983). Academic work. *Review of Educational Research*, **53**(2), 159–199.

Doyle, W. (1988). Work in mathematics classes: The context of students' thinking during instruction. *Educational Psychologist*, **23**, 167–180.

Edwards, L. D. (1999). Odds and evens: mathematical reasoning and informal proof among high school students. *Journal of Mathematical Behavior*, **17**(4), 489–504.

Epp, S. (2009). Proof issues with existential quantification. In: F.-L. Lin, F.-J. Hsieh, G. Hanna, & M. de Villiers (eds.), *Proceedings of ICMI Study 19: Proof and proving in mathematics education* (Vol. 1, pp. 154–159). Taipei, Taiwan: Springer.

Fawcett, H. P. (1938). *The nature of proof (1938 Yearbook of the National Council of Teachers of Mathematics)*. New York: Bureau of Publications, Teachers College, Columbia University.

Fischbein, E., Deri, M., Nello, M. S., & Marino, M. S. (1985). The role of implicit models in solving verbal problems in multiplication and division. *Journal for Research in Mathematics Education*, **16**, 3–17.

Freudenthal, H. (1973). *Mathematics as an educational task*. Dordrecht, Netherlands: Reidel.

Fujita, T. & Jones, K. (2014). Reasoning-and-proving in geometry in school mathematics textbooks in Japan. *International Journal of Educational Research*, **64**, 81–91.

Garuti, R., Boero, P., & Lemut, E. (1998). Cognitive unity of theorems and difficulty of proof. *Proceedings of the 22nd Conference of the International Group for the Psychology of Mathematics Education*, Vol. 2 (pp. 345–352). Stellenbosch, South Africa: University of Stellenbosch.

Gibson, D. (1998). Students' use of diagrams to develop proofs in an introductory real analysis. *Research in Collegiate Mathematics Education*, **2**, 284–307.

Goetting, M. (1995). *The college students' understanding of mathematical proof*. Unpublished doctoral dissertation, University of Maryland, College Park.

Goulding, M., Rowland, T., & Barber, P. (2002). Does it matter? Primary teacher trainees' subject knowledge in mathematics. *British Educational Research Journal*, **28**, 689–704.

Goulding, M. & Suggate, J. (2001). Opening a can of worms: Investigating primary teachers' subject knowledge in mathematics. *Mathematics Education Review*, **13**, 41–44.

Hadas, N., Hershkowitz, R., & Schwarz, B. (2000). The role of contradiction and uncertainty in promoting the need to prove in dynamic geometry environments. *Educational Studies in Mathematics*, **44**, 127–150.

Hagger, H. & McIntyre, D. (2000). What can research tell us about teacher education? *Oxford Review of Education*, **26**, 483–494.

Hanna, G. (1983). *Rigorous proof in mathematics education*. Toronto: OISE Press.

Hanna, G. (1990). Some pedagogical aspects of proof. *Interchange*, **21**(1), 6–13.

Hanna, G. (1995). Challenges to the importance of proof. *For the Learning of Mathematics,* **15**(3), 42–49.
Hanna, G. (2000). Proof, explanation and exploration: An overview. *Educational Studies in Mathematics,* **44**, 5–23.
Hanna, G. & Barbeau, E. (2008). Proofs as bearers of mathematical knowledge. *ZDM—The International Journal on Mathematics Education,* **40**, 345–353.
Hanna, G. & de Villiers, M. (eds.). (2012). *Proof and proving in mathematics education: The 19th ICMI Study.* Dordrecht, The Netherlands: Springer.
Hanna, G. & Jahnke, H. N. (1996). Proof and proving. In: A. Bishop, K. Clements, C. Keitel, J. Kilpatrick, & C. Laborde (eds.), *International handbook of mathematics education* (pp. 877–908). Dordrecht, Netherlands: Kluwer Academic Publishers.
Hanna, G., Jahnke, H. N., & Pulte, H. (eds.). (2010). *Explanation and proof in mathematics: Philosophical and educational perspectives.* New York, NY: Springer.
Harel, G. (1998). Two dual assertions: The first on learning and the second on teaching (or vice versa). *The American Mathematical Monthly,* **105**, 497–507.
Harel, G. (2002). The development of mathematical induction as a proof scheme: A model for DNR-based instruction. In: S. Campbell & R. Zaskis (eds.), *Learning and teaching number theory: Research in cognition and instruction* (pp. 185–212). Norwood, NJ: Ablex Publishing Corporation.
Harel, G. & Sowder, L. (1998). Students' proof schemes: Results from exploratory studies. In: A. H. Schoenfeld, J. Kaput, & E. Dubinsky (eds.), *Research in collegiate mathematics education III* (pp. 234–283). Providence, RI: American Mathematical Society.
Harel, G. & Sowder, L. (2007). Toward comprehensive perspectives on the learning and teaching of proof. In: F. K. Lester (ed.), *Second handbook of research on mathematics teaching and learning* (pp. 805–842). Greenwich, CT: Information Age Publishing.
Harel, G. & Sowder, L. (2009). College instructors' views of students vis-à-vis proof. In: M. Blanton, D. Stylianou, & E. Knuth (eds.), *Teaching proof across the grades: A K-12 perspective* (pp. 275–289). New York, NY: Routledge.
Haylock, D. (with Manning, R.) (2014). *Mathematics explained for primary teachers* (5th edn.). London: SAGE.
Healy, L. & Hoyles, C. (2000). Proof conceptions in algebra. *Journal for Research in Mathematics Education,* **31**(4), 396–428.
Heaton, R. M. (2000). *Teaching mathematics to the new standards: Relearning the dance.* Practitioner inquiry series. New York, NY: Teachers College Press.
Hefendehl-Hebeker, L. (1991). Negative numbers: Obstacles in their evolution from intuitive to intellectual constructs. *For the Learning of Mathematics,* **11**(1), 26–32.
Henningsen, M. & Stein, M. K. (1997). Mathematical tasks and student cognition: Classroom-based factors that support and inhibit high-level mathematical thinking and reasoning. *Journal for Research in Mathematics Education,* **28**, 524–549.
Herbst, P. & Brach, C. (2006). Proving and doing proofs in high school geometry classes: What is it that is going on for students? *Cognition and Instruction,* **24**(1), 73–122.
Hersh, R. (1993). Proving is convincing and explaining. *Educational Studies in Mathematics,* **24**, 389–399.
Hiebert, J., Carpenter, T. P., Fennema, E., Fuson, K., Human, P., Murray, H., Olivier, A., & Wearne, D. (1996). Problem solving as a basis for reform in curriculum and instruction: The case of mathematics. *Educational Researcher,* **25**(4), 12–21.
Hiebert, J., Gallimore, R., Garnier, H., Givvin, K. B., Hollingsworth, H., Jacobs, J., Chui, A. M., Wearne, D., Smith, M., Kersting, N., Manaster, A., Tseng, E., Etterbeek, W., Manaster, C., Gonzales, P.,

& Stigler, J. (2003). *Teaching mathematics in seven countries: Results from the TIMSS 1999 Video Study*. NCES 2003-2013. Washington, DC: US Department of Education, Institute of Education Sciences.
Hodds, M., Alcock, L., & Inglis, M. (2014). Self-explanation training improves proof comprehension. *Journal for Research in Mathematics Education*, **45**, 62–101.
Hoyles, C. & Küchemann, D. (2002). Students' understanding of logical implication. *Educational Studies in Mathematics*, **51**, 193–223.
Iscimen, F. A. (2011). Preservice middle school teachers' beliefs about the place of proof in school mathematics. In: B. Ubuz (ed.), *Proceedings of the 35th Conference of the International Group for the Psychology of Mathematics Education*, Vol. 3 (pp. 57–64). Ankara, Turkey: Middle East Technical University.
Jahnke, H. N. & Wambach, R. (2013). Understanding what a proof is: a classroom-based approach. *ZDM—The International Journal on Mathematics Education*, **45**, 469–482.
Kaput, J. J., Carraher, D. W., & Blanton, M. L. (eds.). (2007). *Algebra in the early grades*. Mahwah, NJ: Erlbaum.
Kazemi, E., Franke, M. L., & Lampert, M. (2009). Developing pedagogies in teacher education to support novice teachers' ability to enact ambitious instruction. In: R. Hunter, B. Bicknell, & T. Burgess (eds.), *Crossing divides: Proceedings of the 32nd Annual Conference of the Mathematics Education Research Group of Australasia*, Vol. 1 (pp. 11–29). Wellington, New Zealand: Mathematics Education Research Group of Australasia.
Kieran, C. (1992). The learning and teaching of school algebra. In: D. A. Grouws (ed.), *Handbook of research on mathematics teaching and learning* (pp. 390–419). New York, NY: Macmillan.
Kieran, C., Pang, J., Schifter, D., & Ng, S.F. (2016). *Early algebra: Research into its nature, its learning, its teaching*. New York, NY: Springer.
Kitcher, P. (1984). *The nature of mathematical knowledge*. New York, NY: Oxford University Press.
Kline, M. (1972). *Mathematical thought: From ancient to modern times*. New York, NY: Oxford University Press.
Knuth, E. J. (2002a). Secondary school mathematics teachers' conceptions of proof. *Journal for Research in Mathematics Education*, **33**(5), 379–405.
Knuth, E. J. (2002b). Teachers' conceptions of proof in the context of secondary school mathematics. *Journal of Mathematics Teacher Education*, **5**(1), 61–88.
Knuth, E., Choppin, J., & Bieda, K. (2009). Middle school students' production of mathematical justifications. In: D. Stylianou, M. Blanton, & E. Knuth (eds.), *Teaching and learning proof across the grades: A K–16 perspective* (pp. 153–170). New York, NY: Routledge.
Knuth, E. J., Choppin, J., Slaughter, M., & Sutherland, J. (2002). Mapping the conceptual terrain of middle school students' competencies in justifying and proving. In: D. S. Mewborn, P. Sztajn, D. Y. White, H. G. Weigel, R. L. Bryant, & K. Nooney (eds.), *Proceedings of the 24th Annual Meeting of the North American Chapter of the International Group for the Psychology of Mathematics Education*, Vol. 4 (pp. 1693–1670). Athens, GA: Clearinghouse for Science, Mathematics, and Environmental Education.
Komatsu, K. & Tsujiyama, Y. (2013). Principles of task design to foster proofs and refutations in mathematical learning: Proof problem with diagram. In: C. Margolinas (ed.), *Task design in mathematics education: Proceedings of ICMI Study 22* (pp. 471–479). Oxford, UK.
Komatsu, K., Tsujiyama, Y., Sakamaki, A., & Koike, N. (2014). Proof problems with diagrams: An opportunity for experiencing proofs and refutations. *For the Learning of Mathematics*, **34**(1), 36–42.
Kondratieva, M. (2011). The promise of interconnecting problems for enriching students' experiences in mathematics. *The Mathematics Enthusiast*, **8**(1–2), 344–382.

Ko, Y. Y. & Knuth, E. J. (2009). Undergraduate mathematics majors' writing performance producing proofs and counterexamples about continuous functions. *Journal of Mathematical Behavior*, **28**(1), 68–77.

Ko, Y. Y. & Knuth, E. J. (2013). Validating proofs and counterexamples across content domains: Practices of importance for mathematics majors. *Journal of Mathematical Behavior*, **32**(1), 20–35.

Krummheuer, G. (1995). The ethnography of argumentation. In: P. Cobb & H. Bauersfeld (eds.), *The emergence of mathematical meaning: Interaction in classroom cultures* (pp. 229–269). Hillsdale, NJ: Lawrence Erlbaum.

Küchemann, D. & Hoyles, C. (2001–2003). *Longitudinal Proof Project* [technical reports for Year 8–10 surveys]. London: Institute of Education. Retrieved May 3, 2009, from: http://www.maths-med.co.uk/ioe-proof/techreps.html

Lakatos, I. (1976). *Proofs and refutations: The logic of mathematical discovery*. Cambridge: Cambridge University Press.

Lamon, S. J. (2001). Presenting and representing: From fractions to rational numbers. In: A. Cuoco (ed.), *The roles of representation in school mathematics* (pp. 146–164). Reston, VA: NCTM.

Lampert, M. (1990). When the problem is not the question and the solution is not the answer: Mathematical knowing and teaching. *American Educational Research Journal*, **27**(1), 29–63.

Lampert, M. (1992). Practices and problems in teaching authentic mathematics. In: F. K. Oser, A. Dick, & J. Patry (eds.), *Effective and responsible teaching: The new synthesis* (pp. 295–314). San Francisco: Jossey-Bass Publishers.

Lampert, M. (2001). *Teaching problems and the problems of teaching*. New Haven, CT: Yale University Press.

Lannin, J., Ellis, A. B., & Elliott, R. (2011). *Essential understandings project: Mathematical reasoning (Gr. K-8)*. Reston, VA: National Council of Teachers of Mathematics.

Larsen, S. & Zandieh, M. (2008). Proofs and refutations in the undergraduate mathematics classroom. *Educational Studies in Mathematics*, **67**(3), 205–216.

Leikin, R. (2010). Learning through teaching through the lens of multiple solution tasks. In: R. Leikin & R. Zazkis (eds.), *Learning through teaching mathematics* (pp. 69–85). Dordrecht, Netherlands: Springer.

Leikin, R. & Levav-Waynberg, A. (2008). Solution spaces of multiple-solution connecting tasks as a mirror of the development of mathematics teachers' knowledge. *Canadian Journal of Science, Mathematics and Technology Education*, **8**(3), 233–251.

Leinhardt, G., Zaslavsky, O., & Stein, M. K. (1990). Functions, graph, and graphing: Tasks, learning, and teaching. *Review of Educational Research*, **60**, 1–64.

Leron, U. & Zaslavsky, O. (2013). Generic proving: reflections on scope and method. *For the Learning of Mathematics*, **33**(3), 24–30.

Leung, I. K. C. & Lew H. (2013). The ability of students and teachers to use counter-examples to justify mathematical propositions: a pilot study in South Korea and Hong Kong. *ZDM—The International Journal on Mathematics Education*, **45**, 91–105.

Linchevski, L. & Williams, J. (1999). Using intuition from everyday life in "filling" the gap in children's extension of their number concept to include the negative numbers. *Educational Studies in Mathematics*, **39**, 131–147.

Lin, P. J. & Tsai, W. H. (2012). Fifth graders' mathematics proofs in classroom contexts. In: T. Y. Tso (ed.), *Proceedings of the 36th Conference of the International Group for the Psychology of Mathematics Education*, Vol. 3 (pp. 139–146). Taipei, Taiwan: National Taiwan Normal University.

Lockwood, E., Ellis, A. B., Dogan, M.F., Williams, C., & Knuth, E. (2012). A framework for mathematicians' example-related activity when exploring and proving mathematical conjectures. In: L. R. Van Zoest, J. J. Lo, & J. L. Kratky (eds.), *Proceedings of the 34th Annual Meeting of the North*

American Chapter of the International Group for the Psychology of Mathematics Education (pp. 151–158). Kalamazoo, MI: Western Michigan University.

Maher, C. A. & Martino, A. M. (1996). The development of the idea of mathematical proof: A 5-year case study. *Journal for Research in Mathematics Education*, **27**, 194–214.

Maher, C. A., Powell, A. B., & Uptegrove, E. B. (eds.). (2010). *Combinatorics and reasoning: Representing, justifying and building isomorphisms*. New York, NY: Springer.

Makar, K., Bakker, A., & Ben-Zvi, D. (2015). Scaffolding norms of argumentation-based inquiry in a primary mathematics classroom. *ZDM—The International Journal on Mathematics Education*, **47**, 1107–1120.

Malara, A. & Navarra, G. (2002). ArAl: a project for an early approach to algebraic thinking. In: A. Rogerson (ed.),*Proceedings of the International Conference 'The Humanistic Renaissance in Mathematics Education'* (pp. 228–233). Palermo, Italy.

Malek, A. & Movshovitz-Hadar, N. (2011). The effect of using transparent pseudo-proofs in linear algebra. *Research in Mathematics Education*, **13**(1), 33–58.

Maloney, A. P., Confrey, J., & Nguyen, K. H. (eds.). (2014). *Learning over time: Learning trajectories in mathematics education*. Charlotte, NC: Information Age Publishing.

Mariotti, M. A. (2000). Introduction to proof: The mediation of a dynamic software environment. *Educational Studies in Mathematics*, **44**, 25–53.

Mariotti, M. A. (2006). Proof and proving in mathematics education. In: A. Gutiérrez & P. Boero (eds.), *Handbook of research on the PME: Past, present and future* (pp. 173–204). Rotterdam, The Netherlands: Sense Publishers.

Mariotti, M. A. (2013). Introducing students to geometric theorems: how the teacher can exploit the semiotic potential of a DGS. *ZDM—The International Journal on Mathematics Education*, **45**, 441–452.

Marrades, R. & Gutiérrez, Á. (2000). Proofs produced by secondary school students learning geometry in a dynamic computer environment. *Educational Studies in Mathematics*, **44**, 87–125.

Martin, W. G. & Harel, G. (1989). Proof frames of preservice elementary teachers. *Journal for Research in Mathematics Education*, **20**(1), 41–51.

Mason, J. (with Burton, L. & Stacey, K.) (1982). *Thinking mathematically*. London: Addison-Wesley.

Mason, J. (1998). Enabling teachers to be real teachers: Necessary levels of awareness and structure of attention. *Journal of Mathematics Teacher Education*, **1**, 243–267.

Mason, J. (2009). Teaching as disciplined enquiry. *Teachers and Teaching: Theory and Practice*, **15**(2), 205–223.

Mason, J. & Klymchuk, S. (2009). *Using counter-examples in calculus*. London: Imperial College Press.

Mason, J. & Pimm, D. (1984). Generic examples: Seeing the general in the particular. *Educational Studies in Mathematics*, **15**, 277–289.

Mason, J., Watson, A., & Zaslavsky, O. (eds.) (2007). The role of mathematics tasks in mathematics teacher education. *Journal of Mathematics Teacher Education* [Special Issue], **10**(4–6), 201–440.

Maybin, J., Mercer, N., & Stierer, B. (1992). Scaffolding in the classroom. In: K. Norman (ed.) *Thinking voices: The work of the National Oracy Project* (pp. 165–195). London: Hodder Education.

McCrory, R. & Stylianides, A. J. (2014). Reasoning-and-proving in mathematics textbooks for prospective elementary teachers. *International Journal of Educational Research*, **64**, 119–131.

Mogetta, C., Olivero, F., & Jones, K. (1999). Proving the motivation to prove in a dynamic geometry environment. *Proceedings of the British Society for Research into Learning Mathematics* (pp. 91–96). Lancaster: St Martin's University College.

Moore, R. C. (1994). Making the transition to formal proof. *Educational Studies in Mathematics*, **27**(3), 249–266.

Morris, A. K. (2002). Mathematical reasoning: adults' ability to make the inductive-deductive distinction. *Cognition and Instruction*, **20**(1), 79–118.

Morris, A. K. (2007). Factors affecting pre-service teachers' evaluations of the validity of students' mathematical arguments in classroom contexts. *Cognition and Instruction*, **25**(4), 479–522.

Morris, A. K. (2009). Representations that enable children to engage in deductive reasoning. In: D. A. Stylianou, M. L. Blanton, & E. J. Knuth (eds.), *Teaching and learning proof across the grades: A K-16 perspective* (pp. 87–101). New York: Routledge.

Moss, J. & London McNab, S. (2011). An approach to geometric and numeric patterning. In: J. Cai & E. Knuth (eds.), *Early algebraization* (pp. 277–301). New York, NY: Springer.

Movshovitz-Hadar, N. (1988). Stimulating presentation of theorems followed by responsive proofs. *For the Learning of Mathematics*, **8**(2), 12–19, 30.

Mullis, I. V. S., Martin, M. O., Foy, P., & Arora, A. (2012). *TIMSS 2011 international results in mathematics*. Boston: Boston College/IEA.

National Council of Teachers of Mathematics [NTCM] (1989). *Curriculum and evaluation standards for school mathematics*. Reston, VA: NTCM.

National Council of Teachers of Mathematics [NCTM] (2000). *Principles and standards for school mathematics*. Reston, VA: NTCM.

National Governors Association Center for Best Practices, & Council of Chief State School Officers [NGA & CCSSO]. (2010). *Common core state standards mathematics*. Washington, DC: NGA & CCSSO.

Nicol, C. (1998). Learning to teach mathematics: Questioning, listening, and responding. *Educational Studies in Mathematics*, **37**(1), 45–66.

O'Connor, M. K. & Michaels, S. (1996). Shifting participant frameworks: Orchestrating thinking practices in group discussion. In: D. Hicks (ed.), *Discourse, learning and schooling* (pp. 63–103). New York, NY: Cambridge University Press.

Otten, S., Males, L. M., & Gilbertson, N. J. (2014). The introduction of proof in secondary geometry textbooks. *International Journal of Educational Research*, **64**, 107–118.

Pedemonte, B. (2007). How can the relationship between argumentation and proof be analysed? *Educational Studies in Mathematics*, **66**, 23–41.

Poincaré, H. (1914/2009). *Science and method* (translated F. Maitland). New York, NY: Cosimo.

Polya, G. (1981). *Mathematical discovery: On understanding, learning, and teaching problem solving*. New York, NY: Wiley Combined Edition.

Prusak, N., Hershkowitz, R., & Schwarz, B. B. (2012). From visual reasoning to logical necessity through argumentative design. *Educational Studies in Mathematics*, **79**, 19–40.

Pyke, C. L. (2003). The use of symbols, words, and diagrams as indicators of mathematical cognition: A causal model. *Journal for Research in Mathematics Education*, **34**(5), 406–432.

Radford, L. (2014). The progressive development of early embodied algebraic thinking. *Mathematics Education Research Journal*, **26**(2), 257–277.

Reid, D. (2002). Conjectures and refutations in grade 5 mathematics. *Journal for Research in Mathematics Education*, **33**(1), 5–29.

Reid, D. (2005). The meaning of proof in mathematics education. In: M. Bosch (ed.), *Proceedings of the 4th Conference of the European Society for Research in Mathematics Education* (pp. 458–468). Sant Feliu de Guixols, Spain. Retrieved December 11, 2011, from: http://ermeweb.free.fr/CERME4/CERME4_WG4.pdf

Reid, D. A. & Knipping, C. (2010). *Proof in mathematics education: research, learning, and teaching*. Rotterdam, Netherlands: Sense Publishers.

Remillard, J. T. (2005). Examining key concepts in research on teachers' use of mathematics curricula. *Review of Educational Research*, **75**(2), 211–246.

Rowland, T. (2002). Generic proofs in number theory. In: S. Campbell & R. Zaskis (eds.), *Learning and teaching number theory: Research in cognition and instruction* (pp. 157–183). Norwood, NJ: Ablex Publishing Corporation.

Rowland, T., Hodgen, J., & Solomon, Y. (eds.). (2015). Mathematics teaching: tales of the unexpected. *Research in Mathematics Education* [Special Issue], **17**(2), 71–147.

Rowling, J. K. (1997). *Harry Potter and the sorcerer's stone.* New York, NY: Scholastic.

Russell, S. J., Schifter, D., & Bastable, V. (2011a). *Connecting arithmetic to algebra.* Portsmouth, NH: Heinemann.

Russell, S. J., Schifter, D., & Bastable, V. (2011b). Developing algebraic thinking in the context of arithmetic. In: J. Cai & E. Knuth (eds.), *Early algebraization: A global dialogue from multiple perspectives* (pp. 43–69). Heidelberg: Springer.

Russell, S. J., Schifter, D., Bastable, V., & Franke, M. (submitted). Bringing early algebra to the elementary classroom: Results of a professional development program for teachers.

Russell, S. J., Schifter, D., Bastable, V., Higgins, T., & Kasman, R. (in press). *Mathematical argument in the elementary classroom: A yearlong focus on the arithmetic operations.* Portsmouth, NH: Heinemann.

Sahin, A. & Kulm, G. (2008). Sixth grade mathematics teachers' intentions and use of probing, guiding, and factual questions. *Journal of Mathematics Teacher Education,* **11**, 221–241.

Samkoff, A., Lai, Y., & Weber, K. (2012). On the different ways that mathematicians use diagrams in proof construction. *Research in Mathematics Education,* **14**(1), 49–67.

Scherrer, J. & Stein, M. K. (2013). Effects of a coding intervention on what teachers learn to notice during whole-group discussion. *Journal of Mathematics Teacher Education,* **16**, 105–124.

Schifter, D. (2009). Representation-based proof in the elementary grades. In: D. A. Stylianou, M. L. Blanton, & E. J. Knuth (eds.), *Teaching and learning proof across the grades: A K-16 perspective* (pp. 71–86). New York: Routledge.

Schifter, D., Monk, S., Russell, S. J., & Bastable, V. (2008). What does understanding the laws of arithmetic mean in the elementary grades? In: J. J. Kaput, D. W. Carraher, & M. L. Blanton (eds.), *Algebra in the early grades* (pp. 413–448). New York, NY: Routledge.

Schoenfeld, A. H. (1985). *Mathematical problem solving.* Orlando, FL: Academic Press.

Schoenfeld, A. H. (1992). Learning to think mathematically: Problem solving, metacognition, and sense making in mathematics. In: D. A. Grouws (ed.), *Handbook of research on mathematics teaching and learning* (pp. 334–370). New York: Macmillan.

Schoenfeld, A. H. (2007). Early algebra as mathematical sense making. In: J. J. Kaput, D. W. Carraher, & M. L. Blanton (eds.), *Algebra in the early grades* (pp. 481–512). Mahwah, NJ: Erlbaum.

Schoenfeld, A. H. (2009). Series editor's foreword: The soul of mathematics. In: D. A. Stylianou, M. L. Blanton, & E. J. Knuth (eds.), *Teaching and learning proof across the grades: A K-16 perspective* (pp. xii–xvi). New York, NY: Routledge.

Schwab, J. J. (1978). Education and the structure of the disciplines. In: J. Westbury & N. J. Wilkof (eds.), *Science, curriculum, and liberal education: Selected Essays* (pp. 229–272). Chicago: University of Chicago Press.

Sears, R. (2012). *An examination of how teachers use curriculum materials for the teaching of proof in high school geometry.* Unpublished doctoral dissertation, University of Missouri, Columbia.

Sears, R. & Chávez, Ó. (2014). Opportunities to engage with proof: the nature of proof tasks in two geometry textbooks and its influence on enacted lessons. *ZDM—The International Journal on Mathematics Education,* **46**, 767–780.

Senk, S. L. (1989). van Hiele levels and achievement in writing geometry proofs. *Journal for Research in Mathematics Education,* **20**, 309–321.

Sfard, A. (1991). On the dual nature of mathematical conceptions: Reflections on processes and objects as different sides of the same coin. *Educational Studies in Mathematics*, **22**, 1–36.

Sfard, A. (2001). There is more to discourse than meets the ears: Looking at thinking as communicating to learn more about mathematical learning. *Educational Studies in Mathematics*, **46**, 13–57.

Sherin, M. G. (2002). When teaching becomes learning. *Cognition and Instruction*, **20**(2), 119–150.

Simon, M. A. (1995). Reconstructing mathematics pedagogy from a constructivist perspective. *Journal for Research in Mathematics Education*, **26**, 114–145.

Simon, M. A. & Blume, G. W. (1996). Justification in the mathematics classroom: A study of prospective elementary teachers. *Journal of Mathematical Behavior*, **15**, 3–31.

Smith, M. S., Silver, E. A., Stein, M. K., Boston, M., & Henningsen, M. A. (2005). *Improving instruction in rational numbers and proportionality: Using cases to transform mathematics teaching and learning, Volume 1*. New York: Teachers College Press.

Sowder, L. & Harel, G. (1998). Types of students' justifications. *The Mathematics Teacher*, **91**(8), 670–675.

Sowder, L. & Harel, G. (2003). Case studies of mathematics majors' proof understanding, construction, and appreciation. *Canadian Journal of Science, Mathematics and Technology Education*, **3**, 251–267.

Stacey, M. K., Chick, H., & Kendal., M. (2004). *The future of the teaching and learning of algebra*. Boston, MA: Kluwer Publications.

Standards & Testing Agency (2014). *Key stage 2 mathematics: Sample questions, mark schemes and commentary for 2016 assessments*. Department for Education. Retrieved August 13, 2015, from: https://www.gov.uk/government/uploads/system/uploads/attachment_data/file/328886/2014_KS2_mathematics_sample_materials.pdf

Steen, L. (1999). Twenty questions about mathematical reasoning. In: L. Stiff (ed.), *Developing mathematical reasoning in grades K-12. 1999 yearbook of the National Council of Teachers of Mathematics* (pp. 270–285). Reston, VA: National Council of Teachers of Mathematics.

Stein, M. K., Engle, R. A., Smith, M. S., & Hughes, E. K. (2008). Orchestrating productive mathematical discussions: Five practices for helping teachers move beyond show and tell. *Mathematical Thinking and Learning*, **10**(4), 313–340.

Stein, M. K., Grover, B., & Henningsen, M. (1996). Building student capacity for mathematical thinking and reasoning: An analysis of mathematical tasks used in reform classrooms. *American Educational Research Journal*, **33**, 455–488.

Stein, M. K., Smith, M. S., Henningsen, M. A., & Silver, E. A. (2000). *Implementing Standards-based mathematics instruction: A casebook for professional development*. New York, NY: Teachers College Press.

Stylianides, A. J. (2005). *Proof and proving in school mathematics instruction: Making the elementary grades part of the equation*. Unpublished doctoral dissertation, University of Michigan, Ann Arbor.

Stylianides, A. J. (2007a). Introducing young children to the role of assumptions in proving. *Mathematical Thinking and Learning*, **9**, 361–385.

Stylianides, A. J. (2007b). Proof and proving in school mathematics. *Journal for Research in Mathematics Education*, **38**, 289–321.

Stylianides, A. J. (2007c). The notion of proof in the context of elementary school mathematics. *Educational Studies in Mathematics*, **65**, 1–20.

Stylianides, A. J. (2009a). Breaking the equation "empirical argument = proof." *Mathematics Teaching*, **213**, 9–14.

Stylianides, A. J. & Ball, D. L. (2008). Understanding and describing mathematical knowledge for teaching: Knowledge about proof for engaging students in the activity of proving. *Journal of Mathematics Teacher Education*, **11**, 307–332.

Stylianides, A. J., Bieda, K. N., & Morselli, F. (2016a). Proof and argumentation in mathematics education research. In: A. Gutiérrez, G. C. Leder, & P. Boero (eds.), *Second handbook of research on the psychology of mathematics education* (pp. 315–351). Rotterdam, The Netherlands: Sense Publishers.

Stylianides, A. J. & Stylianides, G. J. (2008a). Studying the implementation of tasks in classroom settings: High-level mathematics tasks embedded in real-life contexts. *Teaching and Teacher Education*, **24**, 859–875.

Stylianides, A. J. & Stylianides, G. J. (2009a). Proof constructions and evaluations. *Educational Studies in Mathematics*, **72**, 237–253.

Stylianides, A. J. & Stylianides, G. J. (2013). Seeking research-grounded solutions to problems of practice: Classroom-based interventions in mathematics education. *ZDM—The International Journal on Mathematics Education*, **45**(3), 333–341.

Stylianides, A. J. & Stylianides, G. J. (2014a). Impacting positively on students' mathematical problem solving beliefs: An instructional intervention of short duration. *The Journal of Mathematical Behavior*, **33**, 8–29.

Stylianides, A. J., Stylianides, G. J., & Philippou, G. N. (2002). University students' conceptions of empirical proof and proof by counterexample. In: M. Tzekaki (ed.), *Proceedings of the 5th Panellenian Conference on Didactics of Mathematics and Computers in Education* (pp. 150–158). Volos, Greece: University of Thessaly. (In Greek.)

Stylianides, A. J., Stylianides, G. J., & Philippou, G. N. (2004). Undergraduate students' understanding of the contraposition equivalence rule in symbolic and verbal contexts. *Educational Studies in Mathematics*, **55**, 133–162.

Stylianides, G. J. (2008a). An analytic framework of reasoning-and-proving. *For the Learning of Mathematics*, **28**(1), 9–16.

Stylianides, G. J. (2008b). Investigating the guidance offered to teachers in curriculum materials: The case of proof in mathematics. *International Journal of Science and Mathematics Education*, **6**(1), 191–215.

Stylianides, G. J. (2008c). Proof in school mathematics curriculum: A historical perspective. *Mediterranean Journal for Research in Mathematics Education*, 7(1), 23–50.

Stylianides, G. J. (2009b). Reasoning-and-proving in school mathematics textbooks. *Mathematical Thinking and Learning*, **11**, 258–288.

Stylianides, G. J. (2014). Textbook analyses on reasoning-and-proving: Significance and methodological challenges. *International Journal of Educational Research*, **64**, 63–70.

Stylianides, G. J. (2016). *Curricular resources and classroom use: The case of mathematics*. New York: Oxford University Press.

Stylianides, G. J. & Stylianides, A. J. (2008b). Proof in school mathematics: Insights from psychological research into students' ability for deductive reasoning. *Mathematical Thinking and Learning*, **10**(2), 103–133.

Stylianides, G. J. & Stylianides, A. J. (2009b). Facilitating the transition from empirical arguments to proof. *Journal for Research in Mathematics Education*, **40**, 314–352.

Stylianides, G. J. & Stylianides, A. J. (2014b). The role of instructional engineering in reducing the uncertainties of ambitious teaching. *Cognition and Instruction*, **32**(4), 374–415.

Stylianides, G. J. & Stylianides, A. J. (2015). Creating a need for proof. In: E. A. Silver & P. A. Kenney (eds.), *More lessons learned from research*, Vol. 1 (pp. 9–22). Reston, VA: National Council of Teachers of Mathematics.

Stylianides, G. J., Stylianides, A. J., & Philippou, G. N. (2007). Preservice teachers' knowledge of proof by mathematical induction. *Journal of Mathematics Teacher Education*, **10**, 145–166.

Stylianides, G. J., Stylianides, A. J., & Shilling-Traina, L. N. (2013). Prospective teachers' challenges in teaching reasoning-and-proving. *International Journal of Science and Mathematics Education*, **11**, 1463–1490.

Stylianides, G. J., Stylianides, A. J., & Weber, K. (2016b). Research on the teaching and learning of proof: Taking stock and moving forward. In: J. Cai (ed.), *First compendium for research in mathematics education*. Reston, VA: National Council of Teachers of Mathematics. (In press.)

Stylianou, D. A., Blanton, M. L., & Knuth, E. J. (eds.). (2009). *Teaching and learning proof across the grades: A K-16 perspective*. New York, NY: Routledge.

Sullivan, P., Clarke, D., & Clarke, B. (2013). *Teaching with tasks for effective mathematics learning*. New York: Springer.

Sullivan, P. & Mousley, J. (2001). Thinking teaching: Seeing mathematics teachers as active decision makers. In: F. L. Lin & T. J. Cooney (eds.), *Making sense of mathematics teacher education* (pp. 147–164). Dordrecht: Kluwer.

Tall, D. (1999). The cognitive development of proof: Is mathematical proof for all or for some? In: Z. Usiskin (ed.), *Developments in school mathematics education around the world*, Vol. 4 (pp. 117–136). Reston, VA: National Council of Teachers of Mathematics.

Tall, D., Yevdokimov, O., Koichu, B., Whiteley, W., Kondratieva, M., & Cheng, Y. (2012). Cognitive development of proof. In: G. Hanna & M. de Villiers (eds.), *Proof and proving in mathematics education: The 19th ICMI Study* (pp. 13–49). Dordrecht, The Netherlands: Springer.

Tarr, J. E., Chávez, Ó., Reys, R. E., & Reys, B. J. (2006). From the written to the enacted curricula: The intermediary role of middle school mathematics teachers in shaping students' opportunity to learn. *School Science and Mathematics*, **106**(4), 191–201.

Thompson, D. R., Senk, S. L., & Johnson, G. J. (2012). Opportunities to learn reasoning and proof in high school mathematics textbooks. *Journal for Research in Mathematics Education*, **43**, 253–295.

Thompson, I. (2003). Place value: the English disease? In: I. Thompson (ed.), *Enhancing primary mathematics teaching* (pp. 181–190). Maidenhead: Open University Press.

Tirosh, D., Tsamir, P., & Hershkovitz, S. (2008). Insights into children's intuitions of addition, subtraction, multiplication and division. In: A. D. Cockburn & G. Littler (eds.), *Mathematical misconceptions: A guide for primary teachers* (pp. 54–70). London: Sage.

Usiskin, Z. (1987). Resolving the continuing dilemmas in school geometry. In: M. M. Lindquist & A. P. Shulte (eds.), *Learning and teaching geometry, K-12* (pp. 17–31). Reston, VA: National Council of Teachers of Mathematics.

van Es. E. A. & Sherin, M. G. (2002). Learning to notice: Scaffolding new teachers' interpretations of classroom interactions. *Journal of Technology and Teacher Education*, **10**(4), 571–596.

Wagner, S. & Kieran, C. (eds.). (1989). *Research issues in the learning and teaching of algebra* (Volume 4 of *Research agenda for mathematics education*). Reston, VA: National Council of Teachers of Mathematics.

Watson, A. & Chick, H. (2011). Qualities of examples in learning and teaching. *ZDM—The International Journal on Mathematics Education*, **43**, 283–294.

Watson, A. & Mason, J. (2005). *Mathematics as a constructive activity: Learners generating examples*. Mahwah, NJ: Lawrence Erlbaum Associates.

Weber, K. (2002). Beyond proving and explaining: Proofs that justify the use of definitions and axiomatic structures and proofs that illustrate technique. *For the Learning of Mathematics*, **22**(3), 14–17.

Weber, K. (2006). Investigating and teaching the processes used to construct proofs. *Research in Collegiate Mathematics Education*, **6**, 197–232.

Weber, K. (2008). How mathematicians determine if an argument is a valid proof. *Journal for Research in Mathematics Education*, **39**(4), 431–459.

Weber, K. (2010). Proofs that develop insight. *For the Learning of Mathematics*, **30**(1), 32–36.

Weber, K. (2012). Mathematicians' perspectives on their pedagogical practice with respect to proof. *International Journal of Mathematics Education in Science and Technology*, **43**(4), 463–475.

Weber, K. & Alcock, L. (2004). Semantic and syntactic proof productions. *Educational Studies in Mathematics*, **56**(2–3), 209–234.
Weber, K. & Alcock, L. (2009). Proof in advanced mathematics classes: Semantic and syntactic reasoning in the representation system of proof. In: D.A. Stylianou, M.L. Blanton, & E. Knuth (eds.), *Teaching and learning proof across the grades: A K-16 perspective* (pp. 323–338). New York: Routledge.
Wood, D., Bruner, J. S., & Ross, G. (1976). The role of tutoring in problem solving. *Journal of Child Psychology and Psychiatry*, **17**, 89–100.
Wood, T. (1999). Creating a context for argument in mathematics class. *Journal for Research in Mathematics Education*, **30**, 171–191.
Wu, H. (2002). *What is so difficult about the preparation of mathematics teachers?* Unpublished manuscript. Retrieved April 2, 2002, from: http://www.math.berkeley.edu/~wu/
Yackel, E. & Cobb, P. (1996). Sociomathematical norms, argumentation, and autonomy in mathematics. *Journal for Research in Mathematics Education*, **27**, 458–477.
Yackel, E. & Hanna, G. (2003). Reasoning and proof. In: J. Kilpatrick, W. G. Martin, & D. Schifter (eds.), *A research companion to principles and standards for school mathematics* (pp. 22–44). Reston, VA: National Council of Teachers of Mathematics.
Yeager, D. S. & Walton, G. M. (2011). Social-psychological interventions in education: They're not magic. *Review of Educational Research*, **81**, 267–301.
Yu, J. Y. W., Chin, E. T., & Lin, C. J. (2004). Taiwanese junior high school students' understanding about the validity of conditional statements. *International Journal of Science and Mathematics Education*, **2**, 257–285.
Zack, V. (1997). "You have to prove us wrong": Proof at the elementary school level. In: E. Pehkonen (ed.), *Proceedings of the 21st Conference of the International Group for the Psychology of Mathematics Education*, Vol. 4 (pp. 291–298). Lahti, Finland: University of Helsinki.
Zack, V. (1999). Everyday and mathematical language in children's argumentation about proof. *Educational Review*, **51**(2), 129–146.
Zaslavsky, O. (2005). Seizing the opportunity to create uncertainty in learning mathematics. *Educational Studies in Mathematics*, **60**, 297–321.
Zaslavsky, O. (2014). Thinking with and through examples. In: C. Nicol, S. Oesterie, P. Liljedahl, & D. Allan (eds.), *Proceedings of the 38th Conference of the International Group for the Psychology of Mathematics Education*, Vol. 1 (pp. 21–34). Vancouver, Canada.
Zaslavsky, O., Nickerson, S. D., Stylianides, A. J., Kidron, I., & Winicki, G. (2012). The need for proof and proving: mathematical and pedagogical perspectives. In: G. Hanna & M. de Villiers (eds.), *Proof and proving in mathematics education: The 19th ICMI Study* (pp. 215–229). Dordrecht, The Netherlands: Springer.
Zaslavsky, O. & Ron, G. (1998). Students' understanding of the role of counter-examples. In: A. Olivier & K. Newstead (eds.), *Proceedings of the 22nd Annual Meeting of the International Group for the Psychology of Mathematics Education*, Vol. 4 (pp. 225–232). Stellenbosch, South Africa: University of Stellenbosch.
Zazkis, R., Liljedahl, P., & Chernoff, E. J. (2008). The role of examples in forming and refuting generalizations. *ZDM—The International Journal on Mathematics Education*, **40**, 131–141.

AUTHOR INDEX

A

Alcock, L. 10–11, 22, 24, 120, 162
Alexander, R. J. 22
Anderson, J. A. 24, 162
Andriano, V. 132
Arias, A. M. 23
Arora, A. 23, 164–165
Arsac, G. 150
Arzarello, F. 132

B

Bakker, A. 121, 136, 159–160
Balacheff, N. 10–11, 17–18, 30, 83–84, 122, 135, 148, 152, 157
Ball, D. 5, 8–10, 12, 15, 19–23, 34–35, 58–67, 81–97, 114, 119, 137–144, 147, 151, 159, 161, 165
Barbeau, E. 12
Barber, P. 19, 21
Bartolini Bussi, M. G. 12, 28
Bass, H. 8–9, 12, 19, 88, 159
Bastable, V. 86
Bell, A. W. 12, 29
Ben-Zvi, D. 159
Berman, P. W. 86
Bieda, K. N. 5, 19–20, 22–24, 164
Blanton, M. L. 4, 8, 10, 86
Blume, G. W. 21, 121–122
Boero, P. 12, 28, 89, 144
Boston, M. D. 5, 22–24, 164
Brach, C. 32
Brizuela, B. M. 10, 86
Brousseau, G. 149
Brown, S. 24, 162
Bruner, J. 8–9, 13, 36, 68, 151, 160
Buchbinder, O. 10, 19

C

Cai, J. 23, 86, 164
Carpenter, T. P. 9–10, 17, 27, 34, 57, 67, 86, 147, 158
Carraher, D. 10, 86
Cengiz, N. 159
Chávez, O. 5, 24, 157
Chazan, D. 10, 19
Cheng, Y. 7
Chernoff, E. J. 18
Chick, H. 18, 167
Chin, E. T. 10
Choppin, J. 10, 19–20
Christiansen, B. 5, 24, 157
Chui, A. M. 67
Clarke, B. 24, 157
Clarke, D. 24, 157
Cobb, P. 10, 151–152, 159–160
Coe, E. 10, 19
Cohen, D. K. 23, 165
Collopy, R. 32
Confrey, J. 68
Corbett, A. 24, 162
Corey, D. 32

D

Davis, E. A. 7, 23, 165
Davis, P. 7
Davis, R. B. 22, 70
de Millo, R. 152
de Villiers, M. 4, 12, 14, 29, 70
Deri, M. 27, 132, 158
Dewey, J. 13, 19, 67
Dogan, M. F. 12
Doyle, W. 22, 24, 32
Drwenke, J. 23, 24, 165
Durand-Guerrier, V. 150

E

Earnest, D. 10, 86
Edwards, L. D. 19
Elliott, R. 18
Ellis, A. B. 12, 18
Engle, R. A. 22–23, 31–32, 72, 120–121, 150, 159–160
Epp. S. 150
Etterbeek, W. 20, 67

F

Fennema, E. 9, 27, 67
Fischbein, E. 27, 132, 158
Foy, P 23, 164–165
Franke, M. L. 9–10, 17, 27, 34, 57, 86, 147, 158
Freudenthal, H. 14, 87
Fujita, T. 23, 165
Fuson, K. 9, 67

G

Gallimore, R. 20
Gamoran, S. M. 32
Garnier, H. 67
Garuti, R. 12, 28, 89, 144
Gibson, D. 120
Gilbertson, N. J. 23
Givvin, K. 67
Goetting, M. 19, 21
Gonzales, P. 20, 67
Grant, T. J. 159
Grover, B. 32
Gutiérrez, A. 122, 157

H

Hadas, N. 19, 21
Hagger, H. 37
Hanna, G. 4, 8–9, 12, 147, 151–152, 159
Harel, G. 10, 12, 18–19, 21–22, 89, 97, 116, 119, 121, 149, 157, 162
Haylock, D. 27, 132–133
Healy, L. 10, 19
Heaton, R. M. 22
Hefendehl-Hebeker, L. 84
Henningsen, M. A. 22–24, 32, 157, 164
Herbst, P. 32
Hersh, R. 7, 152
Hershkowitz, R. 24, 28
Hershkowitz, S. 132–133
Hiebert, J. 9, 20, 67
Higgins, T. 86
Hodds, M. 24, 162
Hodgen, J. 23
Hollingsworth, H. 67

Hoyles, C. 8, 10, 19–20, 86, 119, 149
Hughes, E. K. 22–23, 31–32, 72, 120–121, 150, 159–160
Human, P. 9, 67

I
Inglis, M. 24, 162

J
Jacobs, J. 67
Jahnke, H. N. 4, 7–10, 13, 20, 24, 33, 67–68, 86, 147, 151, 162
Ji, X 23–24, 165
Jones, K. 23, 30, 165

K
Kaput, J. J. 10, 57, 86
Kasman, R. 86
Kazemi, E. 22, 159–160
Kendal, M. 57
Kersting, N. 20, 67
Kidron, I. 121, 164
Kitcher, P. 7, 70, 84
Kline, K. 159
Kline, M. 84
Klymchuk, S. 148
Knipping, C. 4
Knuth, E. 4, 8, 10, 12, 19–21, 86, 148
Ko, Y. Y. 21, 148
Koedinger, K. 24, 162
Koichu, B. 7
Komatsu, K. 28
Kondratieva, M. 7, 28
Krajik, J. S. 7, 23, 165
Krummheuer, G. 89, 144
Küchemann, D. 10, 19–20, 119, 149
Kulm, G. 89

L
Lai, Y. 120
Lakatos, I. 7
Lamon, S. J. 120
Lampert, M. 9, 15, 17, 22, 30, 37, 68, 151–152
Lannin, J. 18
Larsen, S. 12
Leikin, R. 28
Leinhardt, G. 24
Lemut, E. 12, 28
Leron, U. 17
Lester, F. K. 23, 164
Leung, I. K. C. 148
Levav-Waynberg, A. 28
Levi, L. 9–10, 34, 57, 86, 147
Lew, H. 148
Liljedahl, P. 18
Lin, P. J. 10, 18, 30
Linchevski, L. 85
Lipton, R. 152
Lockwood, E. 12

M
McCrory, R. 24, 163
Mcintyre, D. 37
McKnight, C. 22
Maher, C. A. 10, 19, 34, 120
Makar, K. 159
Malek, A. 17
Males, L. M. 23
Maloney, A. P. 68
Manaster, A. 20, 67
Manaster, C. 20, 67
Manning, R. 27, 132–133
Marino, M. S. 27, 132, 158
Mariotti, M. A. 10, 12–13, 15, 28, 89, 144
Marrades, R. 122, 157
Martin, M. O. 23, 164–165
Martin, W. G. 19, 21, 23
Martino, A. M. 10, 19
Mason, J. 9, 12, 17, 24, 28, 30, 96, 117, 120, 148
Maybin, J. 151, 160
Mercer, N. 151, 160
Michaels, S. 159
Mogetta, C. 30
Moore, R. C. 10, 19
Morris, A. K. 19, 21, 119–120, 149
Mousley, J. 23
Movshovitz-Hadar, A. 8, 17, 135
Movshovitz-Hadar, N. 8, 10, 17, 20, 86
Mullis, I. V. S. 23, 164–165
Murray, H. 9, 67

N
Nello, M. 27, 132, 158
Nguyen, K. H. 68
Ni, Y. 23, 164
Nickerson, S. D. 121, 164
Nicol, C. 159

O
O'Connor, M. K. 159
Olivero, F. 30, 132
Olivier, A. 9, 67
Otten, S. 23

P
Palincsar, A. S. 23
Pedemonte, B. 12
Pelletier, R. 24, 162
Perlis, A. 152
Phillipou, G. 21
Picard, A. 23–24, 165
Pimm, D. 17, 135
Pligge, M. 86
Poincaré, H. 88
Polya, G. 7
Powell, A. B. 34, 120
Prusak, N. 24, 28
Pyke, C. L. 120

R
Reid, D. 4, 10–11, 18–19, 132, 149, 159
Remillard, J. T. 32, 165
Reys, B. J. 23
Reys, R. E. 23
Robutti, O. 132
Ron, G. 148
Ross, G. 151, 160
Rowland, T. 17, 19, 23, 72
Rowling, J. K. 97
Roy, A. R. 165
Russell, S. J. 86
Ruthven, K. 10, 19

S
Sahin, A. 89
Samkoff, A. 120
Scherrer, J. 23, 159
Schifter, D. 86, 120
Schliemann, A. D. 10, 86
Schoenfeld, A. H. 7, 10, 12, 24, 30, 162
Schwab, J. J. 8–9, 13
Schwarz, B. B. 24
Sears, R. 5, 20, 24, 157
Senk, S. L. 10, 20, 23

Sfard, A. 84, 120
Sherin, M. G. 23, 159, 160
Shilling-Traina, L. N. 22
Silver, E. A. 5, 164
Simon, M. A. 21, 68, 121–122
Simpson, A. P. 120
Slaughter, M. 10, 19
Smit, J. 121, 136, 160
Smith, D. O. 165
Smith, M. 5, 20, 22–24, 31–32, 67, 72, 120–121, 150, 159–160, 164
Sowder, L. 10, 12, 18–19, 21–22, 119, 122, 149, 157
Stacey, M. K. 57
Steen, L. 7
Stein, M. K. 22–24, 31–32, 72, 120–121, 150, 157, 159–160, 164
Stierer, B. 151, 160
Stigler, J. 20, 67
Stylianides, A. J. 5, 9–14, 16–17, 19, 21–24, 28, 30–33, 35, 58–68, 72, 88, 91–97, 116–117, 119, 121–122, 133, 136–143, 147, 149–152, 155, 159–160, 162–165
Stylianides, G. J. 5, 7–8, 10–12, 16–17, 19, 21–23, 28, 31–32, 36, 72, 116–117, 119, 121–122, 132, 147, 149–150, 155, 159–160
Stylianou, D. A. 4, 8
Suggate, J. 19
Sullivan, P. 23–24, 157
Sutherland, J. 10, 19

T

Tall, D. 7, 17, 135
Tarr, J. E. 23
Thompson, D. R. 23, 29, 165
Tirosh, D. 132–133
Tsai, W. H. 10, 18, 30
Tsamir, P. 132–133
Tseng, E. 20, 67
Tsujiyama, Y. 28

U

Uptegrove, E. B. 34, 120
Usiskin, Z. 10

V

van Es, E. A. 23, 159

W

Walther, G. 5, 24, 157
Walton, G. M. 162
Wambach, R. 24, 33, 67–68, 162
Watson, A. 18, 20, 24, 30
Wearne, D. 9, 20, 67
Weber, K. 5, 10–12, 22, 24, 119–120, 155, 162
Wegerif, R. 121, 136, 160
Whiteley, W. 7
Williams, C. 12
Williams, J. 85
Winicki, G. 121, 164
Wood, D. 151, 160
Wood, T. 10, 151, 159–160
Wu, H. 88

Y

Yackel, E. 8–10, 151–152
Yeager, D. S. 162
Yevdokimov, O. 7
Yu, J. 10

Z

Zandieh, M. 12
Zaslavsky, O. 10, 17–19, 24, 28, 70, 121, 148, 164
Zazkis, R. 18

SUBJECT INDEX

Only selected mentions of a term are listed in the subject index.

A

addition 34, 44–45, 52–58, 74–85, 130, 132, 140, 160
ad infinitum 54–57, 66, 71, 154
algebra 10, 14, 57, 86, 130, 132, 136, 145, 151, 161
arguments
- categorizations of student 122, 157
- components of 13–14, 40, 148, 160–161
- convincing 12, 152
- definition of 13
- empirical 14, 18–19, 21, 56, 96–97, 116, 119–121, 143–144, 146–150, 154–155, 158, 161
- generic 17–18, 135–136, 151, 154, 156, 158–161
- non-genuine 84, 154, 160
- generated in the episodes 154

argumentation (*see also* modes of argumentation) 12, 89, 144, 159
arithmetic 10, 27, 57, 80, 86–87
- concepts 10, 57, 71, 81, 84–87, 89, 160, 166
- operations 27, 42–43, 57–58, 71, 81–85, 87, 158
- properties (*see also* commutativity) 57, 71

assumptions 14, 32–33, 41, 61–62, 64–72, 88, 117–118, 133
- and their relationship to definitions 69–72

authentic mathematics 9, 31, 68, 166
awareness 83–84, 117
axioms 14, 68, 71
- local 14, 87–88

B

beliefs. *See* teacher

C

calculation 10, 14, 73–89, 133, 140, 147, 154, 156–157, 166
- and its relationship to proving 86–87, 157

Cartesian product problems 91, 113–114, 118, 157
code 105–117, 120, 161
combination problems 91, 95, 118, 157
communication 12, 29, 120
commutativity 42–49, 52–53, 56–57, 69, 82
conjecture 7–8, 12, 14, 18, 30–31, 38–39, 81, 123–152, 156, 158–159, 161
contradiction 12, 14–15, 33–35, 80, 144, 155
contraposition 14, 21
converse 14
conviction. *See* arguments
counterexample 14, 21, 33, 133–134, 136, 143, 148, 150, 154, 158, 161
curricular resources (*see also* textbooks) 5, 23, 162, 164–166
curriculum 8–9, 23, 36, 38, 58, 81, 87, 157
- frameworks 4, 7–9, 20, 23
- developers 5, 10, 21, 24, 163

D

decimal numbers 42, 45–46, 48, 52, 55, 125, 130, 134
definitions (*see also* assumptions) 7, 12, 14, 17, 34, 69–72, 88, 139–140, 144–145, 148–149, 151, 154, 161
didactical break 10
discovery 12, 29
discussion 32, 52, 54, 56, 79, 84, 89, 117, 121, 135–136, 147, 159–160, 164

E

early algebra. *See* algebra
educative curriculum materials (*see also* curricular resources) 165
empirical arguments. *See* arguments
even numbers 17, 34, 88, 136–147
examples (*see also* counterexample) 12, 18–19, 54, 66, 124–129, 132–139, 142, 146–147, 150
explanation 12–13, 29, 84

F

fractions 125–126, 133

G

generalization 12, 18, 21, 31, 57, 70, 119, 123, 134, 136, 143, 148–150, 154, 156, 158, 161–163
generic arguments. *See* arguments
geometry 7–8, 10, 19–20, 68, 86

H

hierarchical classification 70

I

infinity 43, 63, 66
instructional engineering 31–32, 162

instructional intervention 84–85, 89, 150, 160–162
instructional planning 160, 164
instructional reform 5, 165
instructional sequence 71
instructional support 21, 23–24, 163, 165
integers 29–31, 36, 46, 52–53, 55, 57–58, 73, 81, 83–85, 87, 89, 160
intellectual need 89, 97, 116, 119, 121
inverse 14, 21

J
justification 12, 29–31, 34, 39, 52, 64, 84, 117, 132, 136, 143, 149–150, 154–156, 163

K
knowledge (*see also* teacher) 7, 9, 14, 19, 22, 29, 34–35, 55, 57, 67, 80–81, 84–89, 96, 114, 118, 132–133, 136, 145, 149, 151, 160

L
learning residue 70–71, 87, 118
learning trajectory 68, 121–122, 157
local axioms. *See* axioms

M
mathematics as a discipline 8–9, 11, 13, 15, 19, 22, 36, 67–68, 151
misconception 132, 146
modes of argument representation 13–14, 40, 116–122, 145, 160–161
modes of argumentation 13–19, 21, 33–34, 40, 54, 56–57, 65–67, 70–71, 79–80, 96–97, 114, 116, 118–120, 133–135, 145–153, 160–162
models 58–60, 81–85, 87–89, 107, 130–131, 135, 154, 160
modus ponens 14
modus tollens 14
multiples of a number 28–29, 129–130, 132, 135–136
multiplication 29, 31, 123–136

N
natural numbers 30, 34, 97, 127, 130, 132, 134–135, 139–140, 144–145
negative numbers 45–46, 52–62, 81–85, 89, 125, 133, 160
non-genuine arguments. *See* arguments
number sentences 41–66, 69, 71, 81–86, 125–126

O
odd numbers 17, 34, 88, 136–147

P
pattern 2–4, 12, 17–18, 50, 57, 107, 129
pedagogical practices. *See* teaching practices
permutation problems 91, 95, 113, 118, 157
place value 74, 77, 79–81, 87, 143
proof (*see also* proving)
  by contradiction. *See* contradiction
  by contraposition. *See* contraposition
  by counterexample. *See* counterexample
  by exhaustion 14, 118
  by mathematical induction 35
  definition of 13–19
  examples of 15–17, 54, 56, 64–65, 79, 96, 116, 133, 135, 144–145
proving (*see also* proving tasks)
  and its place in classrooms 19–25, 36–37, 86–87, 162–166
  definition of 11–13
  functions (or purposes) of 12–13, 28–29, 84, 156
  importance of 7–10
proving tasks
  and their relationship to proving activity 32–35, 68–70, 87–88, 118–120, 148–149, 153–157
  categorization of 27–31
  characteristics of 32–35
  designing 158
  implementing 158–162
  involving a single case 73–89
  involving infinitely many cases 123–152
  involving multiple but finitely many cases 91–122
  purposes of 28–34, 154–156
  selecting 158
  with ambiguous conditions 41–72

R
reductio ad absurdum 14, 33, 57, 80, 84, 97, 144, 154–155
reflective thinking 67
refutation 12, 29–30, 34, 39, 80, 84, 117, 132–134, 136, 143, 148–150, 154–156, 158, 161, 163
rules of inference 14, 54

S
scaffolding 121, 133, 136, 144, 148, 151
secondary school 1, 10, 19, 21, 24, 86, 145
sense-making 4–5, 9–10, 12, 20, 22, 30–31, 67, 81, 85, 87, 156–157, 159
set
  cardinality 28, 54, 57, 66, 113, 123, 154
  finite 18, 56–57, 66–67, 91, 149, 154–155, 158, 160–161
  infinite 18, 28, 30, 34, 54–55, 119, 123, 134, 136, 143–144, 148–151, 154–156, 158–159, 161–163

of accepted statements 13–14, 55, 64, 84–85, 87–89, 133, 145, 160
skeptic 96, 120, 158, 161
sociomathematical norms 159
subtraction 41–44, 54–55, 58, 80–81
systematic
enumeration 56, 65–67, 96–97, 114
consideration of cases 34, 155
systematization 12, 29

T

tasks (*see also* proving tasks) 24
cognitively demanding 5, 22, 32, 73, 79, 83, 150, 159
teacher
beliefs 21–22, 32, 162–164
knowledge 21–22, 24, 32, 72, 162, 165
role 70–72, 88–89, 120–122, 150–152, 157–162
teacher education 6, 23–24, 37, 162–164, 166
teaching practices 22, 32, 37, 121, 150, 158–160
textbooks (*see also* curricular resources) 9, 23–24, 164–165
theorems 12, 14

U

uncertainty 28, 84
university 1, 19, 24, 37, 150

W

whole numbers 27, 46, 48–49, 52–53, 55–57, 73, 84, 129–130, 145, 158